The Traveller's Guide to

THE CENTRE OF
THE EARTH

ALSO AVAILABLE IN THE TRAVELLER'S GUIDE SERIES
The Traveller's Guide to Infinity
The Traveller's Guide to The Big Bang
The Traveller's Guide to Mars

This edition first published in 2018 by
Palazzo Editions Ltd
15 Church Road
London SW13 9HE

www.palazzoeditions.com

Text & Illustrations © 2018 Palazzo Editions Ltd
Design & Layout © 2018 Palazzo Editions Ltd

A CIP catalogue record for this book is available
from the British Library.
ISBN: 978-1-78675-059-4
Printed and bound in China

Created by Hugh Barker for Palazzo Editions Ltd
Cover art and illustrations by Diane Law

The Traveller's Guide to

THE CENTRE OF THE EARTH

Dougal Jerram
Centre for Earth Evolution and Dynamics,
University of Oslo, Norway/DougalEARTH Ltd., UK

PALAZZO

ABOUT THE AUTHOR

Dougal Jerram is the director of DougalEARTH Ltd. in the UK, and holds an Adjunct Professor II research position at the Centre for Earth Evolution and Dynamics (CEED) at the University of Oslo. He is known as Dr Volcano from his appearances on the BBC, and he has a wide understanding of Earth Sciences and the processes that drive our planet. An enthusiastic earth science communicator, he has been seen on screen as an expert presenter for many television channels including the BBC, CBBC, Channel 4, National Geographic, Discovery, Smithsonian and History channels, and he has published extensively in scientific journals. His other books include; *Introducing Volcanology – A Guide to Hot Rocks*; *Volcanoes of Europe*; and the children's book *Victor the Volcano*.

Contents

The Beagle-Pod

It is no mean feat getting to the centre of the Earth, and in order to do so you will be traveling in your new home-from-home, the Beagle-Pod. Aptly named after the famous voyage of Charles Darwin, the Beagle-Pod, which is the ultimate multicraft equipped with the very latest tunneling technology (see image below), has all of the things at hand that you will need to navigate, explore, and endure your epic adventure.

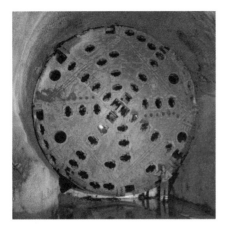

There is a set of key dials, designed in a retro style to give you that *20,000 Leagues Under the Sea* feel, which will establish your location and provide valuable information about the surrounding area. The depth, pressure, temperature, and even the age of the rocks that you are going through will be up on your dials as you embark on your subterranean travels. You can program in aspects of your voyage from a range of pre-determined set of routes. You may also want to program in some of your own special co-ordinates and geological ages to explore. Remember: in your quest for the centre of the Earth, the world is your oyster and, for this purpose, the Beagle-Pod is your key.

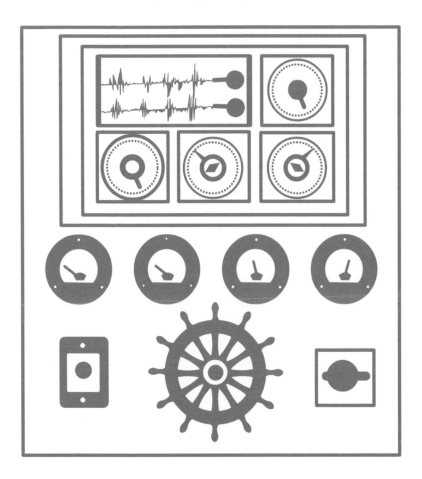

Before departure, you will need to familiarize yourself with the controls of the Beagle-Pod. From top left, here we see (1st row) the seismometer, depth gauge; (2nd row) pressure gauge, speed, temperature; (3rd row) 3D position (in three dials), energy levels; (bottom row) power switch, steering wheel, and power booster.

The Big Picture from Crust to Core

On this quest you will have to go through the layers of the Earth. It will be a long journey with much to see along the way and some great things to explore and learn about. Your first step is to start to become familiar with the layers inside the Earth. Ignoring the atmosphere and the oceans, the layers beneath our feet start with the crust, which can be either continental or oceanic, and which is a relatively thin layer, between about 8 km and 70 km (5-43 miles) thick, making up only 1% of the Earth's volume.

Beneath the crust, the bulk of the planet is made up of the mantle (around 84%) with an average thickness of 2,886 km (1,793 miles). Although the core of the Earth is similarly in thickness to the mantle, at 3,440 km (2137 miles) from the centre to the edge, it only makes up some 15% of the Earth by volume.

You may start your journey on terra firma (from land), the high seas (in the ocean), or even from the very edges of the Earth's atmosphere, as we shall see. Depending on which route you choose, you will experience particular specific aspects of the upper layers of the Earth, whereas for the most part all travellers will experience the majority of what the mantle and core have to offer.

You can of course extend parts of your journey through the upper layers to experience all of its variations. In this guide you will be provided with all the background information you might need as well as highlights to your visit, things to do, and suggestions that will help guide you on your journey all the way to the centre of the Earth, and back out again.

Crust   Core   Mantle

An expanded cross-section of the Earth

How Deep is Deep?

You would imagine that calculating the distance from the Earth's surface to its centre would be fairly straightforward. Work out the diameter of the Earth (maybe by flying round it) and use this to calculate the radius and hey presto, there you have it. Some simple school maths right? Well, it turns out to be a little more complicated than that.

Firstly, the shape of the Earth is not a ball (perfect sphere); it is better described as an "oblate spheroid," that is to say a kind of slightly flattened sphere which is fatter around the middle circumference (equator) and shorter at the top and bottom (the poles). This means that the equatorial radius is around 6,378 km (3,963 miles) whereas the polar radius is 6,357 km (3,950 miles).

When it comes to measuring a planet, we also have to consider the topographic range. This is the amount the topography varies from the deepest trench to the highest mountain. For the Earth this range is 20.4 km (12.7 miles) which also takes into account the oblate shape of the Earth.

As we will see, you have many choices as to how and where you start your journey, but as to how deep you will be going the true answers may surprise you.

North Pole

6,357 km   12,714k m

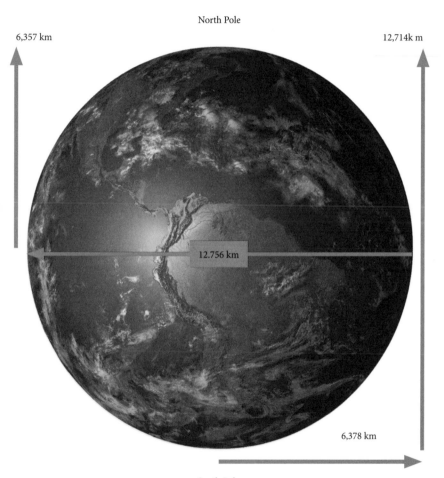

12.756 km

6,378 km

South Pole

The Earth is an oblate spheroid with a polar diameter of 12,714 km
and an equatorial diameter of 12, 756 km

The Earth's Heat Engine

The Earth, and therefore all the rocks on it, originated from a ball of molten material around 4.6 billion years ago. A complicating factor is the fact that if the Earth had simply been left to cool from that time to the present it would have finished cooling and become completely solid a long time ago.

It was William Thomson, later known as Lord Kelvin (famed for the absolute temperature scale that is named after him), that calculated a cooling time for the Earth. If you consider radiation of

The decay of Uranium-238

Key

Half-life units
a = years
d = days
m = minutes

Element names
U = uranium
Th = thorium
Ra = radium
Pa = protactinium
Rn = radon

| Rn-222 | ⬅ | Ra-226 |
| 3.82 d | | 1600 a |

heat alone, the Earth's cooling time should be about 30,000 years. But we know the Earth is still cooling and that it has been doing so for some 4.6 billion years: so why has it not fully cooled? Well, the answers lie in the composition of the rocks that make up the Earth, and more specifically in a group of elements with a particular property. These are the radioactive elements, and the decay of these elements inside the Earth has continually provided a significant additional heat source for the planet.

Elements like uranium and thorium, among many others, are slowly decaying through time. In turn, this decay produces heat at each stage as it is an exothermic (heat giving) reaction. This means the Earth has its own heat engine, and with a half-life of the main uranium isotope U238 of approximately 4.5 billion years, it's a process that can go on for a long, long time.

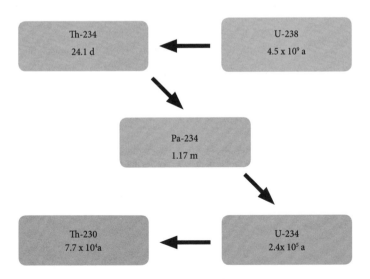

Crust - Continental or Oceanic?

A thin skin on the surface of the planet we call home is the crust beneath our feet. It makes up less than 1% of the Earth's volume but the crust, together with the atmosphere, seas, and oceans on top of it, are all we really know well. This thin layer of rock can be split into two parts, the continental crust (where we reside) and the oceanic crust (beneath our fishy friends).

The continental crust ranges from 25 km to 70 km (15-43 miles) thick and is made up of rocks that are rich in silica and aluminum, as well as all the goodies we have come to know, and to access through the mining process over the years. It forms our mountain ranges, valleys, cliffs and plains, and is constantly changing through weathering and plate movements. The oceanic crust is much thinner

(8-10 km or 5-7 miles) but makes up some 70% of the Earth's surface. It is where all the new crust is formed on the planet. As it is formed underwater, the top part has strange lava flows that develop like a stack of cushions, known as "pillow basalts" (pictured opposite).

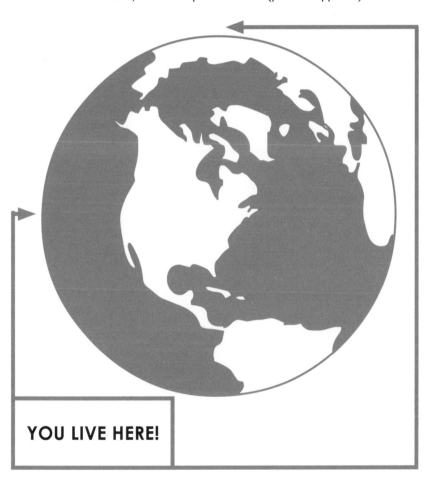

YOU LIVE HERE!

The Mantle

The second main layer of the Earth is the mantle. This is by far the largest layer by volume, making up approximately 84% of the planet. It is broadly broken up into the upper mantle and the lower mantle, but there are other layers and structures within and beyond these layers that make up this colossal mass.

The upper mantle contains a top layer called the lithosphere, which is fixed to the crust above, and below that a region known as

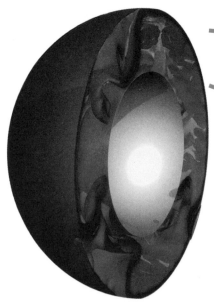

The mantle

the asthenosphere, which behaves in a plastic fashion. The lower mantle, from around 650 km (400 miles) down to the core-mantle boundary at approximately 2,900 km (1,800 miles), is somewhat stronger, and contains rocks rich in magnesium and iron.

You will be spending a significant part of your journey to the centre of the Earth travelling through the mantle, and although large parts of it are relatively homogenous, there are some things to look out for, and you will need to be prepared for a long trip.

As you travel through the mantle, it is not just a case of going straight from top to bottom. On your journey, you can pause to take a look at some of the key layers that are found, such as the low velocity zone, and you can travel along uprising plumes and down-welling slabs as part of the vast engine that drives the plate tectonic movements on the Earth's surface.

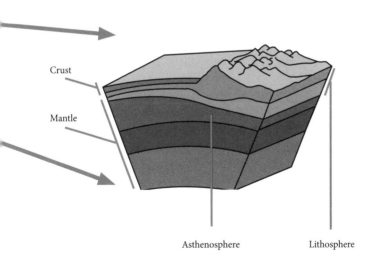

A section of the mantle, going down 2,900 km, with a crust of 8-70km

The Core

Your eventual destination, at least in terms of your quest for the centre of the Earth, the core is often envisaged as a glowing ball of molten rock at the centre of your blue-green planet. It is cited as the cause of our planet's magnetic field, and can cause the poles themselves to flip places.

The core is as mysterious as it is hot, and has the surprising properties of being part liquid (in the outer core) and part solid (in the inner core). The thickness of the core is somewhat similar to, and

The whole planet . . .

just a little larger than that of the mantle at around 3,400 km (2,100 miles) yet it is only 15% by volume of the Earth.

The core of the Earth has been a point of speculation in some iconic (and also some pretty poor) movies over the years, but what will it have in store for you? Well, as we shall be seeing, it is hot and it's under great pressure.

It promises extremes that you will not experience anywhere else, but don't worry: a range of extra safety precautions will certainly be in place on this section of your journey. The core is rich in iron and nickel-iron alloys, and may also contain minor amounts of other metals such as gold and platinum.

Will it be all it's cracked up to be? We will let you be the judge, as you can bring your own sample of the core back, as a memento from your journey to the centre of the Earth.

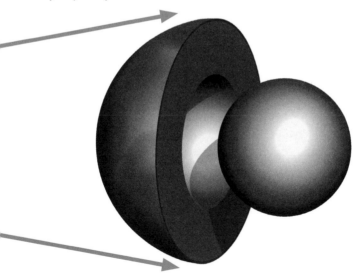

. . . and your final destination, the core

A Convecting Planet

It's true, the world does not stand still, and it's not the day-to-day bustle of life that we are talking about. As we've seen, the inside of the Earth is still generating heat, and as it tries to cool itself down, it moves slowly.

Like a giant, steadily moving lava lamp, the Earth is convecting, moving the heat around and away from the core. We see the results of this manifesting in hot spots and plumes, and the mountain chains and deep trenches that are formed due to plate tectonics.

The crust, like the skin on boiling porridge, wrinkles and buckles above the Earth's convecting interior, and this skin is pierced by the rising plumes. If you could take a 3D view inside the convecting Earth you would see a wondrous vision looking like a lava lamp of hot rising areas, cold descending parts, and mixed zones swirling around.

These drive our surface movements, help build our mountains, and feed our volcanoes and ultimately our atmosphere. As we will see in this guide, you will have many options to explore and drive the pod around the planet's convecting interior, and even use parts of it as your entry and exit points.

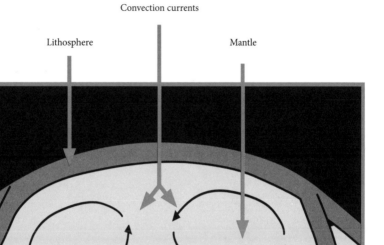

Convection currents

Lithosphere

Mantle

Outer core

Inner core

On your journey you will experience the incredible lava-lamp-style
rotation of convection currents in the mantle of the earth

The Lithospheric Lid

You can think of the lithospheric lid as the coat or the clothes that cover the Earth's hot interior. The planet's crusts, oceanic and continental, are in fact attached to the upper part of the mantle.

The combined "solid" layer that consists of crust and mantle is called the lithosphere, and this rests on top of the asthenosphere and the deeper mantle below. The lithospheric lid marks the parts

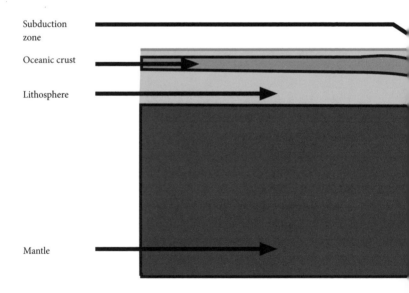

Subduction zone

Oceanic crust

Lithosphere

Mantle

Diagram of a subduction zone

of the Earth that are involved in the slow, steady movement of the plates on top of the asthenosphere.

The thickness of this important layer varies: it can be between 50 km and 120 km, depending on whether it is linked to oceanic or continental crust. At the mid ocean ridges it is no thicker than the oceanic crust itself, and the base of the lithosphere is marked by a zone of partially molten "slow" mantle, where seismic waves are affected and where melt within the Earth can be generated.

It also contains another important layer marking the contact between crust and mantle, called the Moho, which we will be visiting shortly in this guide.

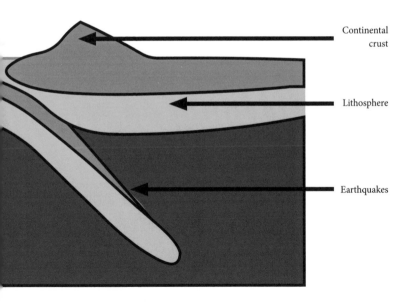

Continental crust

Lithosphere

Earthquakes

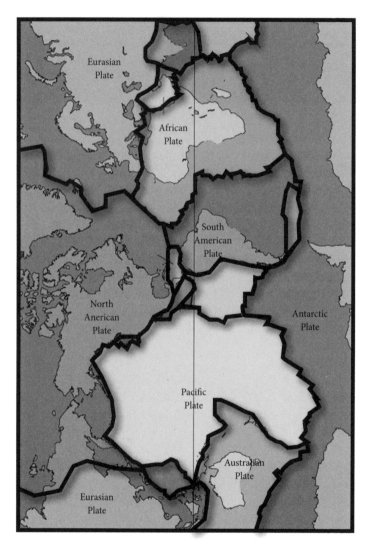

Map of the plates of the world, with some of the major plates labelled

Plate Tectonics

The Earth's surface is made up of relatively rigid regions known as "plates" which are the solid parts which ride around on the convecting planet as it tries to cool. These can be made up of either oceanic or continental crust, and are bound by key contact zones, which can be seen on the Earth's surface as expressions of mountain chains, trenches and ridges.

They owe their discovery to the theories of continental drift and seafloor spreading (which we will explore over subsequent pages). These theories have been combined relatively recently and, along with an understanding of the Earth's convection, they provide us with a clearer picture of the large-scale movement of the Earth's surface through plate tectonics.

These plates are active today, and have been through much of Earth's history. They are responsible for creating many of the phenomena that have been responsible for the rock formations that make up the Earth's crust.

You will be able to see these structures and how they are driven from deep in the mantle, as you take your journey. A striking relationship that you will be able to explore is the correlation between the position of the plate boundaries and the location of the majority of the Earth's volcanoes and earthquakes, as we will be seeing later in this guide.

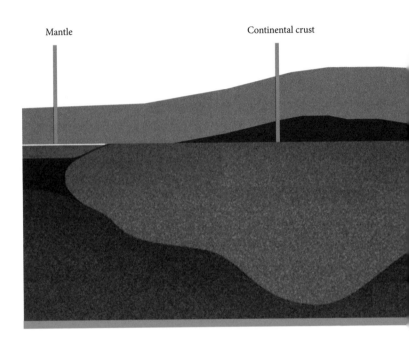

Mantle

Continental crust

Discovering the Moho

A special layer that lies approximately 7-35 km (4.5-21 miles) below the Earth's surface marks out the transition from the crust to the mantle. This is known as the Mohorovičić discontinuity, or "Moho," for short. You will see that this layer is shallow beneath the oceanic crust and deeper beneath the continental crust, and you will notice the properties of the rock you are going through change across this layer. The Moho was discovered by a Croatian geophysicist called Andrija Mohorovičić (1857-1936). Armed with some of the most up-

The Moho Layer from the surface down to 50 km (31 miles)

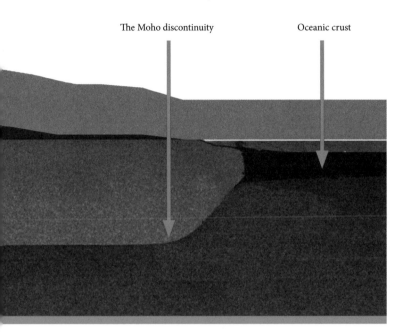

The Moho discontinuity

Oceanic crust

to-date seismic instruments of the period in which he was working, Mohorovičić observed that some seismic waves from an earthquake were arriving much quicker than expected at the instruments.

He concluded that when an earthquake occurred within the uppermost layer of the Earth (the crust), it sent out seismic waves, some of which passed through a lower "faster" layer (the mantle). In doing so he had effectively defined the boundary between the crust and mantle. This layer, whose existence was later confirmed by other studies, has become known as the Mohorovičić discontinuity after this great pioneer of modern seismology.

A Tale Of Two Cores

We have not one but two: one is made of liquid and the other one is solid. This is a riddle that refers to the Earth's core (or should we say cores?) But just how did we discover the whole truth about the centre of the Earth?

In 1906, the British geologist Richard Oldham made some observations about the way seismic waves from an earthquake passed through the Earth. He realized that they must be being deflected by a dense body within the Earth, and in coming to this conclusion he established the existence of the Earth's core. Soon afterwards he published his landmark paper "The Constitution of the Interior of the Earth as Revealed by Earthquakes," but his observations up to this point had not revealed whether the core was liquid or solid. This left Oldham and others to conjecture that it might be very dense and made of iron.

As we were unable at this stage to explore at great depths within the Earth (as you will be doing on your expedition), the debate continued about the exact nature of the core, its depth, and just what this dense body within the Earth actually was.

However, the seismic waves that travelled through the Earth's core were starting to reveal that there must be liquid present. No S waves were transmitted though, so an entirely liquid core was still a possibility at this stage. Some were moving to the idea the core was a dense liquid mass, while others were in stout opposition that it was a dense solid.

People of Note: Richard Oldham

In his twenties, Richard Oldham worked with the Geological Survey of India in the Himalayas. His detailed report on the 1897 Assam earthquake included a description of the Chedrang fault, which had dramatic levels of uplift that reached up to 10.6 metres (35 feet) and reported accelerations of the ground that had exceeded the Earth's gravitational acceleration. He identified the separate arrivals of P waves, S waves, and surface waves on seismograms.

Inge Lehmann

Another major discovery about the nature and structure of the Earth's core was made by the Danish scientist Inge Lehmann. She undertook a close study of data from a New Zealand earthquake in 1929, and discovered that the seismic waves that did make it through the core (P waves) were themselves being deflected by another body within the core, the "inner core." This finally revealed that there must indeed be two cores, a solid inner core and a liquid outer core.

The Alphabet of Earth's Layers

The naming of the layers of the Earth is something that has evolved through time and with different discoveries. The most convenient and obvious method would seem to be to start with A and continue through the alphabet as you come across different layers. This labelling scheme has indeed been used to help define the density distribution within the Earth, with the crust being indicated by the letter A and the inner core ending up with the letter G.

This was the labelling structure used by the New Zealand geophysicist Keith Bullen (1906-1976), who was interested in the Earth's density distribution. His layers were ordered from A to G as in the diagram opposite. These labels are based on the main density shifts in the Earth's structure. This layering definition has not quite stood the test of time as it is insufficiently complex.

However, the use of the letter D for the lower mantle has been given a bit of a lifeline with the discovery of a complex layer within the lower mantle, leading scientists to adopt D' (D prime) for the main part of the lower mantle and D'' (D double prime) for the layer immediately in contact with the core-mantle boundary.

Certain irregularities in this boundary zone are thought to be caused by a novel phase of a mineral called perovskite named post-perovskite, in which seismic waves are also effected.

So as you go down, remember your ABCs!

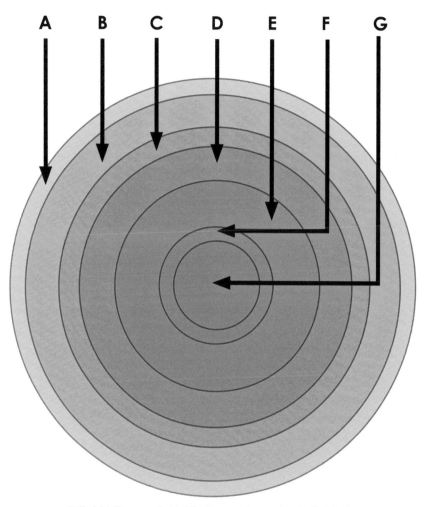

Bullen's labelling started with A for the crust; then continued with B for the upper mantle; C for the transition zone; D for the lower mantle; E for the outer core; G for the inner core; and F for a transitional layer between E and G.

Switching the Poles

Did you know that north has not always been north? In fact it has been south many times, and the South Pole has been the north. This sounds strange but at periods in the Earth's history the magnetic alignment of the poles has swapped round. These are known as polar reversals and they are due to complex interrelationships between thermodynamics, the fluid motions of the outer core, and the evolving magnetic field. But, in basic terms, the north arrow on your compass would be facing south if you were in a polar reversal.

Over the last 20 million years the poles have been reversing every 200,000-300,000 years on average, though the last time it happened was some 780,000 years ago. This strongly suggests that we might be due another one soon. These polar reversals can be very important in the rock record. Volcanic rocks like lava flows contain lots of iron, and when they crystallize, iron-rich minerals can keep a record of the polar conditions at that time.

As we look back through Earth's history, these polar reversals can be used to track time: the surface of the ocean floors become a record of how the plates are moving, as each new batch of ocean crust is formed from lavas which record the alignment of the poles. Also, those traveling along the Earth's axis will note that the position of the magnetic poles is not exactly the same as the location of the true poles, with the magnetic poles having drifted steadily through time.

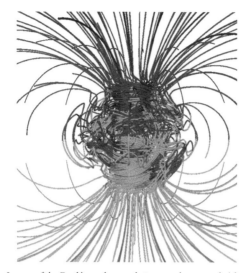

Images of the Earth's geodynamo between polar reversals (above)
and during polar reversals (below)

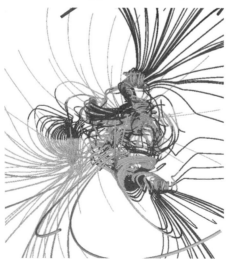

Drifting Continents

Alfred Wegener first proposed the theory of continental drift in 1912. He envisaged the top layers of the Earth drifting slowly above a liquid core. The idea had been touched on a long time earlier by Abraham Ortelius in 1596, who noted the similarity between the outlines of the continents and suggested that the Americas were "torn away from Europe and Africa . . ." The notion of drifting continents was the seed that gave birth to the modern-day theory of plate tectonics.

The fossil record supports and gives credence to the theories of continental drift and plate tectonics. Probably the most compelling part of the continental jigsaw is the match up of Africa with South America. A journey through the rocks on either side of these continents, confirms a match in space and time with regard to the period in which the continents were still together.

Supporters of this theory had a tough time initially and Wegener didn't to see his thoughts fully vindicated. A British geologist, Arthur Holmes, came up with the idea that a convecting planet might drive this process in the late 1920s, and in the 50s and 60s our growing understanding of seafloor spreading saw the final pieces of the puzzle slot into place.

The continents have indeed drifted, and still are drifting. There is evidence of this motion in the rock record, in the patterns on the ocean floor, and in the giant plate tectonic system that moves and juggles our planet's surface around like the the wrinkles in the skin of a wise old man's face.

Wegener was one of those wise men and his theory marked the beginning of the age of earth science that started putting the pieces of the jigsaw back together.

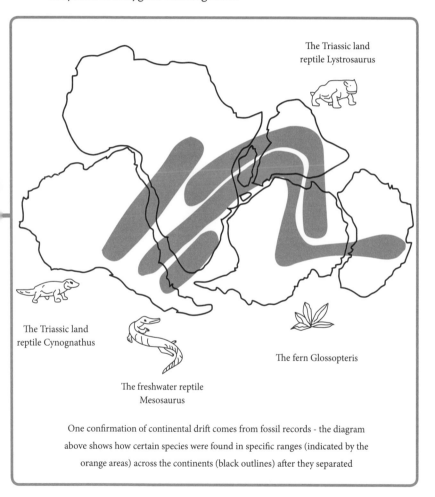

The Triassic land reptile Lystrosaurus

The Triassic land reptile Cynognathus

The freshwater reptile Mesosaurus

The fern Glossopteris

One confirmation of continental drift comes from fossil records - the diagram above shows how certain species were found in specific ranges (indicated by the orange areas) across the continents (black outlines) after they separated

Switching Magnetic Stripes

The world would look a completely different place if you could put on special glasses that showed you different aspects of your physical surroundings. Everyone wonders about x-ray glasses, but if you could see the Earth through magnetic spectacles you would see a remarkable feature. All along the ocean floor as far as the eyes can see you would find strange stripes.

This phenomenon is there because of the changing of the magnetic poles and the near constant production of new crust at

the mid-ocean ridges. Each time the Earth's magnetic poles shift, the new crust that forms takes on the new polarity, and when you look with magnetic vision you see this change as you move from young crust at the ridge, to older crust further away. First proposed by British scientists Vine and Mathews in 1963, this theory became known as the Vine-Matthews-Morley hypothesis, a name which also recognizes the work of Canadian geologist Lawrence Morley, who had independently come up with the idea.

The Spreading Seafloor

This phenomenon was also a vital observation when it came to the notion of seafloor spreading. This is a theory that was pioneered in the 1950s by the former US Navy captain, Harry Hess: he suggested that the Earth's crust must be moving laterally away from the long, volcanically active mid-oceanic ridges.

The truth of this suggestion helps underpin the theory of plate tectonics. In the image, the age of the oceanic lithosphere is indicated by the shading. The darker the area of ocean, the older the lithosphere, with the age range varying from recently formed parts of the ocean floor (at the ridges) to those formed some 180-200 million years ago.

The Types of Plate Boundary

The many wonders of our surface topography owe their existence to the types and dynamics of the plate boundaries. They control where we have mountains, where volcanoes and earthquakes are located, and also provide you with some different options as to where you can start your journey into the upper parts of the Earth's crust and mantle. Boundaries where the crust is moving apart are known as "divergent plate boundaries" or "rifts," Here, new crust is created by magmatism and the asthenosphere reaches up close to the surface.

"Convergent plate boundaries" are those where one plate is subducted under another, which pushes up the crust to make mountains. These are particularly interesting areas to investigate as there is a thicker section of the Earth's crust to explore and you also have the option of moving deeper into the crust along the area where the plate is moving down.

Other forms of plate margin include "transform plate boundaries" where the plates rub side to side. This creates zones like the San Andreas Fault and can lead to some very shaky earthquakes.

Getting Started

The beginning is the most important part of the work.

Plato

Getting Started

The world is a big place, and what we see at the surface is merely a thin layer that covers up a wealth of information beneath. You have chosen to take a journey to the centre of the Earth, but there is a great deal to consider before you set off: not just the things you might want to see on the way, but also a plan for where you want to set out from and where you want to pop out at the other end of your subterranean expedition.

When planning your route you have a variety of exciting opportunities, and in principle the world is your oyster. However, there are a few recommended routes and itineraries which are available to help guide the overwhelmed or unsure traveller on their quest for the centre and their journey back to the surface.

On a map of the globe you can pick any entry point and exit point that you would like, but on the following pages we will explore some options that provide ease of access or allow you to go that extra mile to achieve some other milestones.

We will give you some clues as to how you might delve down into the planet. These are only a few from some of the many choices available, but may help steer you in the right direction.

There are as many different routes to the centre of the Earth as there are travellers making the journey

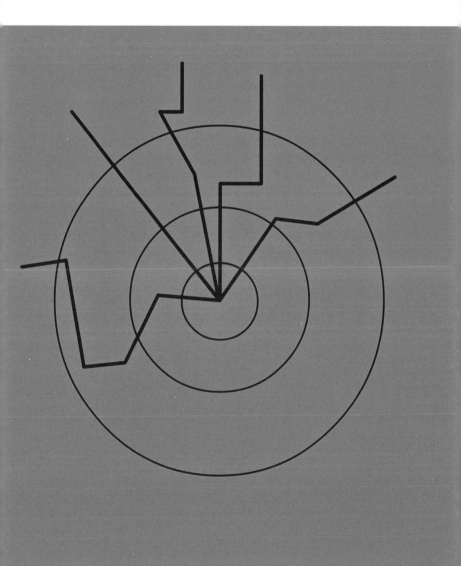

The Classic Volcanic Routes

Drilling into the centre of the Earth requires an entry point, and what could be more appropriate than heading down the throat of a volcano? In some mythologies they are known as the "Gateway to Hell", but in reality they provide a quick route into the planet's interior. In the classic book *Journey to the Centre of the Earth*, Jules Verne has this adventurers descending into the

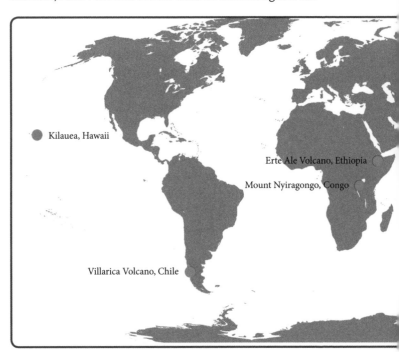

Kilauea, Hawaii

Erte Ale Volcano, Ethiopia

Mount Nyiragongo, Congo

Villarica Volcano, Chile

Earth through Snaefels, on Iceland. In fact there are a number of better volcanic locations on Iceland and even more if you scan the globe for the locations of the active volcanoes.

Some of the best volcanic entries are found in the Earth's active lava lakes. These volcanoes are constantly active, but calm enough for a safe entry into the planet's interior. There are only a modest number of volcanoes that fit this bill, some of which have had active lava lakes for many decades. The oldest and most classic of these lava lakes is the volcano Erta Ale in Ethiopia. This is the original "gateway to hell," as it is named by the local Afar people. A rather more relaxed entry can be made down Villarrica Volcano in Chile from the spa town of Pucon near its base. The lava lakes of Kilauea, Hawaii, and Mount Erebus in Antarctica offer two extreme environments. Whereas the lava lakes of Vanuatu, Mount Nyiragongo in the African Congo, and Masaya Volcano in Nicaragua, offer somewhat more tropical volcanic routes to the core. Whichever volcano you choose, you will be making a spectacular, and rather hot, entry into the Earth.

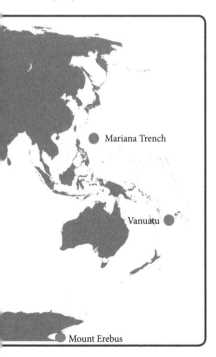

Through the Ocean Deeps

A popular way to enter the Earth's crust is at the deepest point in our oceans. This is known as the Challenger Deep and is found along the Mariana Trench in the Pacific Ocean (see map on previous page). This is a point where one part of the ocean's crust is being subducted beneath another.

Before reaching the ocean floor at Challenger Deep, you will have travelled nearly 11,000 metres (36,000 ft) through the ocean. At this point you will already be experiencing pressures of just over 1,000 bars and temperatures between one and four degrees

In 1960, the bathyscaphe *Trieste,* a precursor of the Beagle-Pod, became the first manned craft to reach the bottom of the Mariana rench

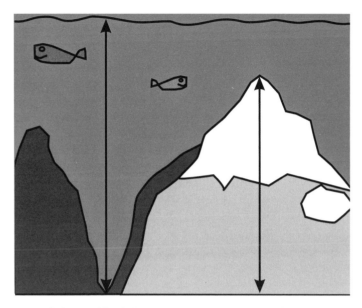

The bottom of the Challenger Deep is 11,000 metres (36,000 feet) below
the surface. By contrast, Mount Everest is 8,848 metres high (29,002 feet)
making it 2,152 metres (7,100 feet) shorter

centigrade, although the protective shield of the Beagle-Pod will
be keeping you perfectly safe at all times. Some people mistakenly
believe that the Mariana Trench provides the shortest route to the
Earth's core. However, this is not the case due to fact that the Earth
is not completely spherical.

The shape of the Earth (an oblate spheroid) means that the poles
are nearer to the core than the equator, which means that the very
deepest parts of the Arctic Ocean can be as much as 13 kilometers
closer to the core than the Challenger Deep. After entry from this
point, you can follow the colder subducting slab down towards the
core as part of the Earth's convecting interior.

The Longest Entry

It may also be tempting to take the longest possible route from the Earth's surface to its centre. So, presumably, to the top of Mount Everest, the highest point above sea level, you must go.

But it's not as simple as that, in fact it's not at the heady heights of Mount Everest that you would start your journey, but in a much hotter place, closer to the equator. In the same way that the deepest trench in the ocean is not necessarily the closest place to the centre of the Earth, its highest mountain is not the furthest place from the planet's core.

The place that holds the record for the farthest distance from the Earth's centre is in fact a volcano. Its name is Chimborazo, and it is located in Ecuador at some 6,384 kilometers (3,966.8 miles) from the centre of the Earth. That is a good 2,168m (7,113 ft) more than the distance from the centre to the summit of Mount Everest.

With the oblate spheroid shape of the Earth meaning that the equatorial radius is greater than the polar radius of the planet, it is the mountains at or near the equator which are further away from the core than those in mountain chains which lie on other latitudes.

So it's Chimborazo Volcano in Ecuador for the true aficionados who want to take the longest route through the interior of the planet, although Everest remains a popular starting point as it is so well known and has its own particular glamour.

Chimborazo Volcano, the closest place on the surface of
the Earth to the sun, pictured from the West at sunset

Pole to Pole

Imagine the Earth as a giant cherry on a cocktail stick, with the stick going through both poles, and that gives you an image of the Earth's axis. Travelling through the Earth's axis is a popular route for travellers within the planet as the intrepid explorer gets to bag both poles along the way.

The journey is around 12,714 km (7,100 miles) with the polar radius of the Earth being approximately 6,357 km (3,950 miles). This is shorter than an equator to equator journey which would be about 31 km (19 miles) longer due to the Earth's shape.

The excitement of the icy launch and arrival as well as the strange compass readings as you head down through the invisible magnetic field all add to the enjoyment of this route. The most common path people take is from North to South Pole, probably because this is classically the way our maps are orientated, with north at the top. So starting from the north seems logical to our traditional way of thinking.

There is the added bonus of coming out at the South Pole where there is a permanent base. Here you can grab a cup of tea when you pop out, and there is an airstrip that can fly you back home afterwards. Look out for polar bears as you get ready for launch in the north and of course penguins when you finally emerge in the south.

Destination Anarctica

Catching All the Layers

If you head down to the centre of the Earth from the level that we live on, you will either tunnel straight through the land you walk on, or take a journey through the seas and oceans to their bottom and beyond. This seems the most normal approach, with the choice of entry and exit being the real question. But some might notice that with this route you miss a number of key layers.

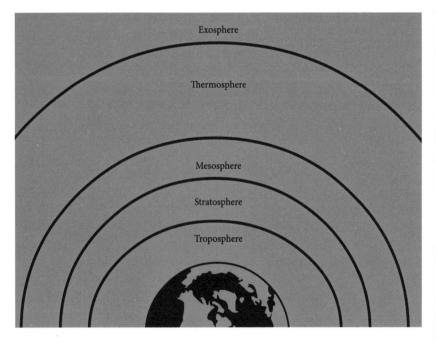

Exosphere

Thermosphere

Mesosphere

Stratosphere

Troposphere

The atmosphere seen from the International Space Station

The planet does not just stop at its surface or at sea level. We have an extensive atmosphere that also makes up some of the layers of the Earth. Some people choose to take a route which captures every possible layer of the Earth from the edge of space right through the middle of the inner core. With this route you need to be aware of the main layers of the atmosphere as these are the added extra that you will experience. The outermost part of the Earth's atmosphere is known as the exosphere and lies 90-10,000 km (428-6213 miles) above our heads. Next comes the thermosphere (85-690 km or 52-428 miles) where you find the aurora and which normally define as the location of the transition between the Earth's atmosphere and outer space (the Kármán line at 100km or 62 miles).

The mesosphere (50-85 km or 31-52 miles) is where most meteors turn into shooting stars. The stratosphere, home to the ozone layer, is wedged between the troposphere and the mesosphere. Finally the troposphere, where we live, contains three quarters of the Earth's atmosphere. It varies in thickness from around 17 km (11 miles) at the equator to as thin as 7 km (4 miles) at the poles.

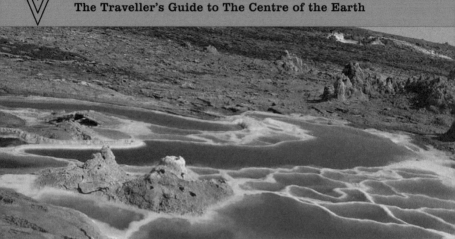

The Lowest & Hottest Place on Earth

If you like extremes then here is an entry place you might want to consider, one of the lowest places (below sea level on the land surface) and the hottest inhabited place on Earth, with surface temperatures that can reach as high as 50 degrees centigrade. At a mine in Dallol in Ehiopia, a record was set when the average temperature (including day and night and both summer and winter) was 35 degrees centigrade.

Much of the area is below sea level with the lowest part at −155 m (−509 feet): you can even fly below sea level here. It's not only a record-breaking entry point for your journey, but the moonscape that is found around the volcano of Dallol, where acid pools and sulphurous yellow crystals adorn the rocky landscape, provides a vivid backdrop for your journey. You can explore the salty layers here as well as look at how the volcanoes pierce their way through the salt flats. But don't hang around, though, the hydrothermal fluids here can have a ph of less than 1; this means they are acids that can dissolve almost everything in their path.

Things to Do

The more we do, the more we can do; the more busy we are, the more leisure we have.

Dag Hammarskjold

Bag the Seven Continents

The beauty of the Beagle-Pod is that you can go absolutely anywhere in it, since it is fully equipped with aquatic elements, and can fly and tunnel as well as travel over the land. In this respect it is the most all-encompassing form of transport ever devised.

Many of those who are heading to the core also want to notch up a number of other feats, and one achievement that many aspire to is going to all seven continents: North America, South America, Europe, Africa, Asia, Australia, and Antarctica. So there is a lot of

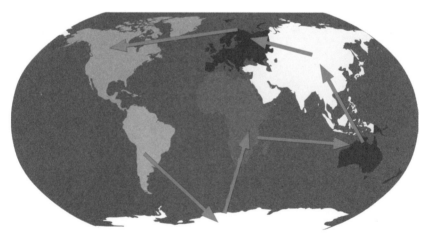

You need to choose a starting point, a direction, and a route
to follow to take in all of the continents

traveling to do, but it can all be done underground. You can decide on some of the key bits of rock real-estate that you want to collect from each continent as proof of your having bagged the seven continents. Popular options include the opal mines of Australia, amethyst layers in South America, the original gold rush seams in North America, and more besides.

Your choices are many. The true collectors and baggers among you will also want to get the full array of oceans into the mix; Pacific, Indian, Arctic, Atlantic . . . Remember that if you aiming for a full list of continents and oceans before heading down to the core, it's going to be a busy collecting trip, with rock shopping available at many destinations around the world. So you will need to allow for the extra time necessary when planning your adventure.

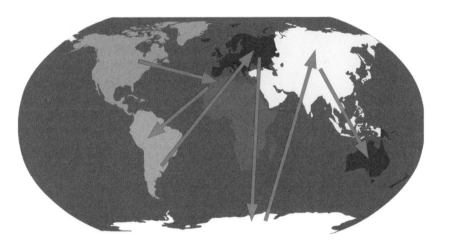

The zig-zag wanderer option

Diamond Hunting Beneath the Cratons

As you travel within the Earth's lithospheric mantle (the upper part attached to the crust), you can take a detour to the areas that lay beneath the oldest parts of the Earth's crust, known as the cratons. These are the locations for exciting "Diamond Hunt" excursions on which you can try to find the birthplace of the hardest material on the Earth.

You will need to travel around 140-300 km (87-190 miles) beneath the cratons, as you are looking for conditions of between 45-60 kilobars in pressure and temperatures in excess of 900 degrees centigrade (1,650 degrees Fahrenheit).

Depending on where you entered the Earth, you would be looking to find diamonds beneath the African, Canadian or Russian cratons, but you may also be lucky enough to find

them under India, Australia, and Brazil. They are brought to the Earth's surface through volcanic pipes known as kimberlites. These strange volcanic beasts are able to erupt material from deep within the Earth, transporting the diamonds quickly to the surface before they are able to break down to natural carbon.

If you are not having much luck finding a diamond, why not hunt out a kimberlite pipe? Then you can travel up and down its broken and twisted structure, and imagine what it would be like to be one of the lucky diamonds plucked from the depths and sent up into the world above.

Hunting for Fossils

As you delve through the sedimentary layers of the various basins that you are traveling through, you can put on your tin hat, get out the geological hammer and go hunting for fossils.

You will need to find a good layer, rich in dead beasts and, depending on the time zone, you will be able to search for the key types of critters that lived in that particular time zone.

It is most common to find marine fossils such as dead shells, bivalves, and brachiopods. A common marine fossil is the curly ammonite which is a great favourite, along with the much older trilobite. A shark's tooth is always another good find, bringing back memories of the film *Jaws*. But the most popular find is that of a dinosaur, and none more than big old T-Rex itself. These are much harder to find as these animals lived mostly on the continents and as such have less chance of being preserved.

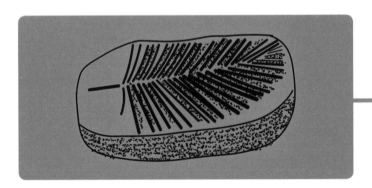

What can be even more intriguing is finding a trace fossil. This is where an animal has left a trace of its movements but the main fossil is not there. Don't feel short-changed if you find one of these, as you may be witnessing some of the earliest life forms with Pre-Cambrian trace fossils. You may also be able to stand in the footsteps of giants if you find a dinosaur footprint.

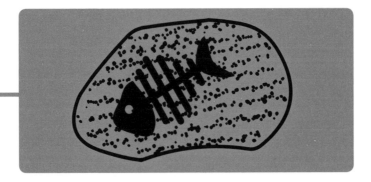

Touching the Rifting Plates

For those doing the classic volcanic entry route in Iceland, there is a hidden treasure in the cold lake waters that you can visit on your way down. Silfra Lake in Iceland (pictured below) stands on the Mid-Atlantic Ridge and because of this you can literally drive your Beagle-Pod along an underwater cavern that runs through the lake. The passageway marks one of the rift faults that form as the plates move apart along the constructive plate boundary. The water is so clean and pure here that you can see for an extraordinary distance, which makes for an eerie perspective.

Here we see the author of this book, at the rift. On the left hand side of the image is the Eurasian Plate, while on the right hand side you can see the North American Plate.

You can pop on your dry suit – the temperature of the lake is barely above freezing – and venture out into the ice-cold, crystal clear waters. The passage, with cliffs either side, represents the divide between the Eurasian and American Plates. At one point you can even put your hands on either side of the underwater rift, as though you are literally pushing the plates apart yourself. The rift is separating at roughly the same rate as your fingernail is growing, so you will not see it moving, but geologically that is still quite fast.

The Ridge Run & the Ocean Factory

They scar the ocean floor, swinging and winding around like a big zip running around the globe. When you look at a map of the Earth with the water removed, the oceanic ridges are by far the longest ranges of mountains on the planet, and for the most part they are still completely unexplored.

The Mid-Atlantic Ridge runs almost the complete length of the planet from north to south, the Pacific Ocean is battle-scarred by a number of ridge segments and the Indian Ridge splits into three routes. The "ridge run," as it is known, is the journey that bags the whole length of this global structure. These ridges are the locations

On this image the rifts in the plates are marked in white. The
Mid-Atlantic Ridge stands out as one of the the longest.

The Remotely Operated Vehicle *Hercules* collecting basaltic rock in the vicinity of the Mid-Atlantic rift on the 2005 Lost City expedition

where new crust is made – the ocean factory – which feeds brand-new ocean crust to Earth, and they are constantly on the move. As you move away from the ridges the crust gets older and older, as shown by the invisible magnetic stripes that mark the sea floor.

Even if you don't choose the full ridge run, many people like to explore sections of the ridges in different parts of the ocean floor before heading deeper into the mantle.

Along the ridge, you are as close as you can get on the surface of the planet to the mantle. In some cases the axis of the mid-ocean ridge is broken or offset laterally by transform (sideways) faults, and parts of the mantle can actually be exposed.

Wander through the Hollows

The ground beneath our feet is not always solid and, as you ride through the shallow parts of the Earth, you have an opportunity to explore its many caves and caverns.

You can head for limestone rocks which are renowned for having many interlinking cave networks, with wonderful natural forms of cave art. Drip stones, that form as mineral-rich water evaporates

in the caves, produce stalactites (which grow downwards from the roofs of the caves) as well as stalagmites (which start on the floor and point upwards). Sometimes these meet in the middle and form columns. You might also be able to submerge yourself in underground lakes and rivers as you navigate your way between the caverns. A slightly more unusual but equally fascinating set of the Earth's hollows are found around volcanoes. Beneath the sides of a volcano where lava flows pour down, you may be lucky enough to find spectacular lava tubes. These represent passageways where lava flowed underground to feed eruptions at the surface, which have now partly drained and solidified resulting in meandering volcanic hollows. As you travel along them you can imagine you are riding a river of glowing magma.

A very special excursion can be made to the most spectacular cave of all: the Crystal Cave in Mexico, which lies beneath a silver mine. This is a site where giant crystals of gypsum have been growing for thousands of years from a volcanically heated hydrothermal liquid. Beautiful clear crystals of selenite (a pure form of gypsum) grow up to 13 metres (43 feet) long, in a criss-cross pattern. Unsurprisingly this hollow also has the nickname of "Superman's Cave."

The author exploring the Crystal Cave

A Primordial Lake Under Antarctica

For those who are taking the pole-to-pole route, you will get a wonderful opportunity to visit a very special place deep under the ice on the continent Antarctica.

Of the many lakes trapped under the ice, Lake Vostok is the largest. The water, which was originally isolated many thousands, and possibly millions of years ago, is thought to be home to prehistoric life forms. These are likely to be microorganisms so will not be easy to see, but keep your eyes peeled as any discovery in this murky primordial soup is likely to be a first.

Scientists had long theorized that water must be present under the pressure exerted by the sheer volume of ice. The existence of this particular lake was only confirmed in 1993 after many years of speculation. This is also the site for the longest ever cored section through ice, a section of 3,768 m (12,400 ft) which was drilled in 2012 by Russian scientists.

The lake surface itself is actually 500 m (1,640 ft) below sea level, and lies beneath ice cover that is 4 km (2.5 miles) thick. It's a dark and mysterious place. This sealed fossil lake is thought to represent conditions that might be present under the icy oceans which are known to exist in other parts of the solar system such as the moons Europa (Jupiter) and Enceladus (Saturn). This has led some to speculate that this ancient hidden lake might hold the key to understanding life on other planets.

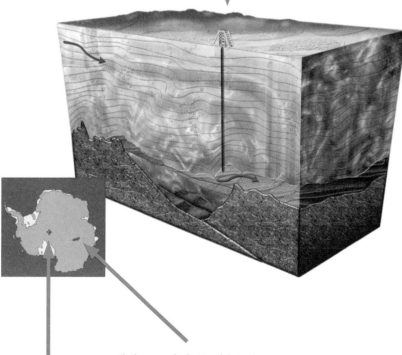

The cored section which reaches down 3,768 m (12,400 feet) to the water below

The location of Lake Vostok beneath the ice of Antarctica

The South Pole

Geyser from the Bottom Up

Normally we only see them from the top when they send gallons of boiling water gushing into the air. A geyser is a hot jet of water and steam which periodically erupts from the ground around hot springs in volcanic terrains. For those choosing the classic volcanic entry in Iceland, you will have the opportunity to see these wonderful natural phenomena from the bottom up. From this vantage point you will see where the heat and energy comes from to flash boil the ground water, and witness a rather unique view of an erupting geyser.

The Great Geysir in Iceland is the daddy of all geysers, and gave them their name. These days it erupts periodically, but in the 19th century it was very active until an earthquake partly curtailed its activity. Over the years, people used the ingenious addition of soap to persuade Great Geysir to erupt again, but more earthquakes in 2000 re-invigorated it naturally. However, its eruptions are still somewhat irregular. The nearby Strokkur geyser is the buddy of Great Geysir and this one erupts fairly regularly, pleasing the many crowds that gather to see it.

Other great geyser locations include Old Faithful and Steamboat in Yellowstone, USA; El Tatio geyser field in Chile; the Valley of Geysers in Kamchatka, Russia; and the Taupo Volcanic Zone in New Zealand. If you want, you can try adding some soap to the water yourself, to see if you can get a particular geyser to erupt.

The Strokkur geyser in Iceland

Riding the Ocean Currents

You may be entering the Earth through a number of different vantage points to explore its inner mysteries, but if you choose an underwater entry you will have the added bonus of being able to explore the world's ocean depths. Deep, dark, and

mysterious, many parts of the oceans are relatively unknown, so why not add a detour to follow the ocean currents around the sea floor and marvel at this submarine kingdom?

For those who are going into the Earth down its deep ocean trenches, or who are already following the trail of the seafloor ridges, riding the ocean currents is another way to explore some of the Earth's massive forces.

These giant movements of water circumnavigate the globe in a winding pathway that can hug the ocean floor or move great currents along the surface waters, and are an essential part of the lifeblood of the planet. For many the currents provide a popular way to get home when they pop up from the Earth's core, allowing the Beagle-Pod to be swept along and watching the world, or more correctly, the ocean go by.

Key currents to look out for are: the Bengulela (along the west coast of Africa); the Humboldt Current along the west coast of South America; the deep polar currents; and the North Atlantic drift which keeps the waters around Europe ice-free in winter.

The main ocean currents

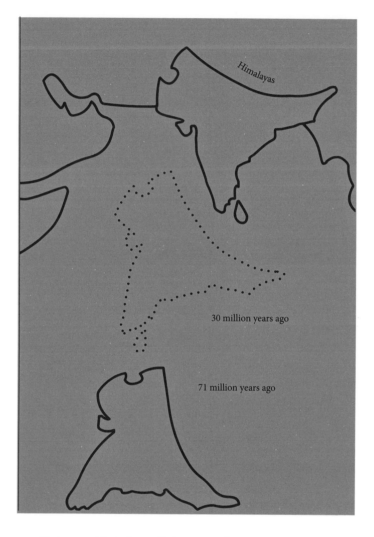

Himalayas

30 million years ago

71 million years ago

The journey of the landmass of India over 71 million years as it crashes into
the Eurasian landmass and continues to push up the Himalayas.

Explore Inside the Himalayas

If you like things big, then nothing comes bigger than the Himalayan mountain chain. If you open up your old-school atlas, or take a turn through the Google Earth or NASA globes, then this colossal range of rocky real estate stands out like a sore thumb. The Himalayas, and its associated range the Karakoram, are host to the fourteen eight-thousander mountains – independent peaks over 8,000 m or 26,500 feet tall – including the biggest, Mount Everest. The colossal structures weave and wind along in an arc some 2,900 km (1,800 miles) long, separating the plains of India from the Tibetan Plateau.

The mountains themselves are the product of larger things happening to the planet, as India is crashing up into Asia in a plate tectonic collision zone. The Himalayas are the bent up bumpers and mangled bonnet of this gigantic car crash.

They form such a large chain that many people like to visit them on their route towards the centre of the planet, and to take some time to explore the roots of the mountain chain. If you choose to set out on your journey from this area, you will find many a metamorphosed rock, and you can even cross from one plate to another by going downwards instead of sideways, as the Indian Plate has been subducted beneath the Asian Plate.

Lighting up the Ring of Fire

The edges of the largest ocean on Earth, the Pacific, where the ocean meets the land, are home to a vast circle of volcanoes known as the "Ring of Fire." Plate tectonic margins at the Pacific edges result in volcano-filled mountains and islands marking a horseshoe shaped ring that runs some 40,000 km (25,000 miles) in length. You can go clockwise from New Zealand or anticlockwise from South America, and even end up at the bubbling lava lake of Mount Erebus in Antarctica, which is also a popular entry point. A fun thing to do as you circumnavigate around the ring of fire is to see how many volcanoes you can bag en route.

Use the Beagle-Pod to explore their structure, but be careful as you could strike magma. Some of the highlights on an anticlockwise journey include: Villarrica Volcano, one of Chile's most active volcanoes, and a must-see in South America; the Cascades range of North America, home to Crater Lake and Mount St. Helens (which erupted in 1980); the valley of a thousand smokes in Alaska; and the volcanoes of the Kamchatka Peninsula, Russia, which also provide a hot and cold experience. The masterful Mount Fuji towers over Tokyo. Mount Pinatubo in the Philippines is another which has erupted as recently as 1991. Lake Toba in Indonesia is famed for its supervolcanic eruption, estimated to have occurred about 74,000 years ago. And arriving at New Zealand you encounter the steaming White Island Volcano and the Taupo Volcanic Zone.

Over 450 volcanoes (around 75% of the world's active or dormant volcanoes) can be drilled through in a subterranean journey around the Ring of Fire.

What to Look Out For

The more we do, the more we can do; the more busy we are, the more leisure we have.

Dag Hammarskjold

Seismic Waves

These are the waves of energy that pass through the Earth: they can be caused by earthquakes, but also by other events such as impacts and even man-made explosions. They mainly occur as two types (P and S waves) and are instrumental in our understanding of the structure of the Earth. You may have seen them being recorded as squiggly lines on seismographs. On your journey, remember that you can use the on-board seismograph in your Beagle-Pod to monitor earthquakes that are happening in real time around the globe.

The way in which the seismic waves travel through the Earth provides a window into its internal structure. Importantly, S waves do not travel through liquids and so these helped us to determine that the outer core of the Earth is liquid.

Also, the way in which these waves get deflected means that there are shadow areas around the planet where no P or S waves are recorded when there is an earthquake event. If you are precisely on the opposite side of the globe from an earthquake you will only be able to register P waves. You can use our real-time record of earthquake events to see if you are indeed in the shadow zone or opposite the earthquake that has happened, using the readings on your seismograph.

We also use the information from several seismic stations around the Earth to locate the true 3D location of the earthquake, known as its epicentre.

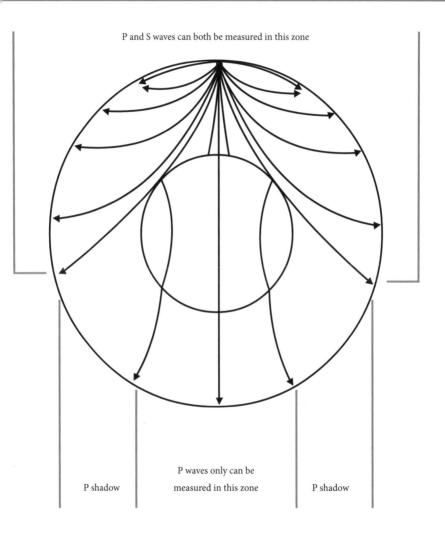

P and S waves can both be measured in this zone

P shadow

P waves only can be
measured in this zone

P shadow

Only P waves travel through the Earth's core. Both P and S waves are refracted
as they travel through the Earth and its core. This is because of the increase
in density and pressure as you approach the centre of the Earth.

The Magnetic Earth

As the outer core is liquid, it does not sit still but swirls around with the spinning motion of the planet. As the molten metal moves round relative to the other parts of the Earth, it creates a giant magnet. Just like the pattern made by iron filings on a piece of paper with a magnet underneath, the Earth's magnet generates large curved fields from pole to pole.

This invisible field actually extends out into space and protects the planet from some harmful particles. It deflects much of the solar wind, which contains charged particles that would otherwise be harmful as they would strip away the Earth's ozone layer. The field also extends within the planet.

Although we don't use classic compasses much these days, most of our gadgets take advantage of the Earth's magnetic field through small internal compasses. Your phone, your GPS and the aircraft you fly in all rely on the Earth's magnetic field.

Historically, man has used the fact that the Earth has a magnetic field to forge expeditions of discovery and adventure. As you descend into the deep, you have a large classic mariner's spherical compass installed in pride of place in the Beagle-Pod, so you can see how the magnetic Earth changes with depth. What will it be like when you get to the centre?

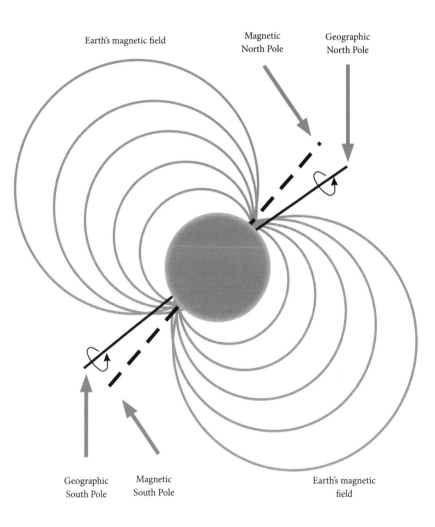

Earth's magnetic field

Magnetic
North Pole

Geographic
North Pole

Geographic
South Pole

Magnetic
South Pole

Earth's magnetic
field

Diagram of the planet's magnetic field and its poles

Where Has All the Magma Gone?

Contrary to a popular misconception, the inside of the Earth is not a ball of molten magma. For the most part it is solid, bar the outer core. It is only in rare conditions that you can get hot rocks to melt. So where has all the magma gone?

Well the trick is that a mix of pressure, heat, and fluid, and the different ways that these can combine, gives us magma, and ultimately leads to the volcanoes we know and love.

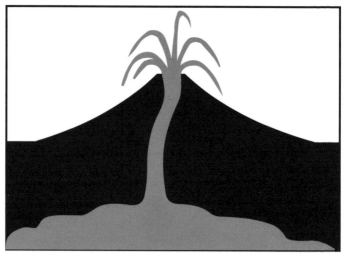

A volcano erupts: the magma passes from the magma chamber up through the vent and erupts through the crater. The volcano acquires extra layers of rock on its surface as the lava cools and solidifies.

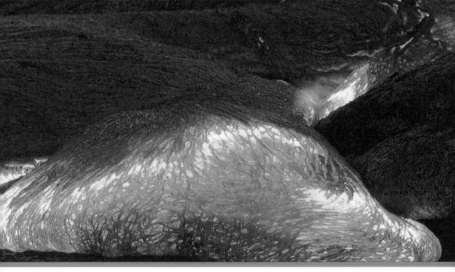

Lava (seen cooling here in Hawaii) is the external form of magma

As we go deeper into the Earth, we get hotter due to what is known as the geothermal gradient, and there must be a point at which we get so hot we can melt rocks. But that's where pressure comes in, because the melting point of rocks (called the solidus) increases with pressure. In ordinary circumstances, these two lines do not normally cross. But the fact we have volcanoes at the surface must mean that there are exceptions.

Indeed there are . . . If you can add fluid to hot rocks you can lower their melting point, and at subduction zones this is what happens, and volcanoes pop up all along these zones. If you can make rocks rise rapidly, they lose pressure faster than they are cooling which means they melt. The same applies to the ocean ridges, where the mantle rises and melts forming the new crust. Finally, take a blowtorch to add heat, like a plume or hot spot (as found for instance in Hawaii) and you can cause melting too.

Simple really!

The Comet's Lair

Deep within the layers of rock on the Yucatán Peninsula in Mexico, there is a hidden secret that dates back to around 66 million years ago. A giant bowl structure, with its centre close to the town of Chicxulub, marks an impact crater made from a massive asteroid or comet strike (the Chicxulub Crater).

This was a significant period in geological time as it represents the change from the Cretaceous to the Paleogene period, known as the End-Cretaceous mass extinction. This was an important event in the history of our planet (particularly if you were a big dinosaur).

Some 16% of all marine families, 47% of all genera, and an estimated 71-81% of all species became extinct, including our beloved dinosaurs, when something huge happened to the planet.

The Chicxulub impact from outer space is implicated in shuffling the big beasts off this mortal coil, and evidenced by this gigantic crater as well as a spike in iridium (a rare metal similar to platinum and an element found in abundance

A near miss: the Great Comet of 1861 passed within 32 million km (20 million miles) of the Earth and was visible to the naked eye for 3 months

in space, but not on Earth) in rocks around the world at this time.

Recent work suggests that it was more likely a comet that struck, as the amount of iridium found is much less than would have been caused if it was a big rocky asteroid. Also, the Deccan Traps, a massive volcanic event, was occuring at the same time and is thought to also have contributed to the Earth crisis.

Imagine, you could be at the site where a giant comet struck the Earth 66 million years ago, with BIG consequences. You can look out for samples of the affected rocks from the impact with "shocked" quartz, tektites (glassy, partly melted samples), and also see if you can find the iridium anomaly within the sediment layers.

Pangea and Gondwana

The constantly moving Earth and the phenomenon of plate tectonics mean that in Earth's past, the continents found themselves together as "supercontinents" such as Pangea and Gondwana (pictured below as they started to break up into our modern day continents). Plate reconstructions that use the matching geology on the continents, as well at the stripes of the ocean floor and the wandering poles, all help to build up a picture of the formation and break-up of these vast landmasses.

There are many online animations of the position of the various plates through time. A fun activity is to drill through the rocks either side of two continents that were once together to see how they match up. For example, you can do this in Namibia in Africa and southern Brazil in South America. If you tune your Beagle-Pod correctly you will be able to see exactly the same sequences of rock, in continents that are now thousands of kilometers apart.

Dangers and Warnings

The most dangerous thing in the world is to try to leap a chasm in two jumps

David Lloyd George

The Centre Of The Earthquake

As you travel through the upper parts of the Earth you will potentially have to cross areas where earthquakes are a risk. These occur in key areas of the map of the world, usually along plate boundaries. You may have chosen to enter the Earth through one of the subduction trenches, where one plate moves beneath another. This is in many ways a good option as you can use the downward motion of the plate towards the mantle. However, these are also areas of earthquake activity.

Earthquake damage at the epicentre of the 2010 Chile earthquake.
There were 358,214 earthquakes recorded between 1963 and 1998

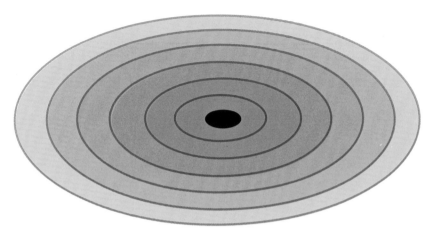

It is advisable to minimize the time you spend in such regions to avoid the danger of being caught up in the centre of an earthquake. The "epicentre" of the earthquake is the point at the Earth's surface directly above the earthquake and the "hypocentre" or "focus" of the earthquake is the exact position of it within the Earth.

Experiencing an earthquake from within the Earth is quite different to doing so on the surface, as your plane of reference is unclear. You will be able to see what type of movement has occurred and the location of the quake with the geophysical instruments within the safety of the Beagle-Pod.

As there are, at any one time, earthquakes happening all over the world, you are highly likely to be able to see some of the recordings of them with the sensitive instruments. Earthquakes can be triggered by the injection of fluids, so it is not recommended that you make a discharge of any sort from the Beagle-Pod during your descent through a subduction zone.

Drilling Into a Magma Chamber

Although it's actually quite hard to find, you might just drill into magma during your journey. We are clearly prepared to meet molten material as we go from the solid lower mantle into the outer liquid core, but what about the chances of a magmatic encounter in the upper parts of the planet?

We know magma exists in small minute pockets within the low velocity zone, but it is in the magma chambers, the magma sheets and conduits that magma uses to rise to the surface and erupt, that we can encounter a sudden change in our surroundings.

As magma cools, parts of the surface become
volcanic rock, floating on the liquid below

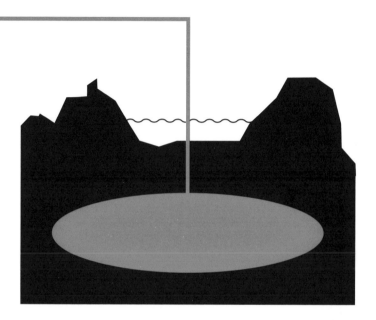

Clearly where you are moving around near the mid-ocean ridges, around a plume or hot spot, or in the areas above a subduction zone, you are at risk of a magma encounter. For the most part you will notice a change in movement and behaviour of the pod, a bit like turbulence in a plane. But be careful: strike at an active volcano and you could get caught up in an eruption, and instead of a journey to the centre you will be on a quick ride back to the surface.

Surprisingly, man has actually drilled into magma before, once intentionally and once by mistake. When the 1959 Kilauea Iki eruption had stopped, it created a massive lava lake. As this cooled, scientists carried out experiments that involved drilling from the cooled lake surface into the molten magma beneath. While in Iceland, explorers looking for geothermal energy once got more than they bargained for when they actually drilled into molten rock.

Feeling Hot, Hot, Hot!

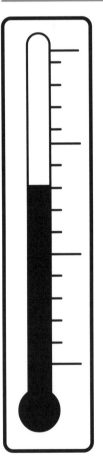

Just how hot will it get as we go down to the core? The amount the temperature goes up with depth is known as the geothermal gradient, and in general it averages around 25 degrees centigrade per kilometre (40 degrees per mile) in the Earth's crust, but can range from 10-50 degrees C/km (16-80 degrees per mile.) It can be more in areas where there is excess heat such as the spreading centres and hot spots or plumes, and a little colder in old crust like cratons.

As you get deeper the gradient changes to about 1-2 degrees C/km (2-3 degrees per mile.) So you will see heat increase relatively rapidly as you go through the crust and upper parts of the Lithosphere, but it will slow down with depth. There is another increase in geothermal gradient as you approach the core-mantle boundary and then it slows off again.

The base of the mantle is over 4,000 degrees centigrade and the inner core can get up to over 6,000 degrees centigrade. This makes the centre of the Earth about the same temperature as the surface of the Sun.

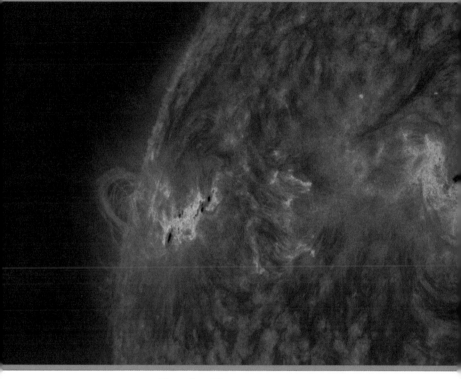

The surface of the sun

degrees Fahrenheit). This makes the centre of the Earth about the same temperature as the surface of the Sun. Which is hot, hot, hot. Estimates for the temperature at the centre of the Sun are in the region of 15,000 degrees centigrade (27,000 degrees Fahrenheit), which is more than twice the maximum temperature you should experience on your trip. As you descend, you can watch the temperature increase on the dials in the pod. These are designed in a classic style to give you that retro *Journey to the Centre of the Earth* feel, and a chart of the actual geotherm that you experience can be printed off as a souvenir for your journey's return pack.

Under Pressure

There is no doubt about it, you will be feeling the pressure as you travel into the planet's interior. Soon there will be a lot of rock above your head and it will be pushing down with force. As you travel deeper you will see the outside pressure indicated on the pressure gauge on the dials and charts wall within your Beagle-Pod. This should gradually rise as you go deeper, but you may see it go up and down if you are exploring areas where you need to travel around laterally.

The pressure gradient, the amount pressure increases, is around 1 GPa or 10 kbar per 35-40 km (22-25 miles) within the Earth's crust. This increases a little more within the mantle, and jumps up in pressure as we enter the mostly metallic core. For comparison, the tyres on your car are around 2 bars (200,000 pascal) and the bottom of the deepest part of the ocean is 1,000 times that of the atmosphere.

At the very centre of the Earth the pressure is around 365 GPa (365,000,000,000 pascal or 3,650,000 bars). Given that the standard atmospheric pressure is ~100,000 pascals, this makes it some 3.65 million times higher.

Are you feeling the force?

Rock Recipes

The geologist takes up the history of the Earth at the point where the archaeologist leaves it, and carries it further back into remote antiquity.

Bal Gangadhar Tilak

The Earth Cookbook

As with many a journey, it is fascinating to explore the local recipes and specialities on offer that we may not have experienced before. With a journey to the centre of the Earth, you are faced with a variety of local compositions of the material that makes up the

The pebbles on a beach can tell us a huge amount about the local varieties of rock

planet, a sort of "Earth's cookbook," in which you sample the many varieties that are on offer. They all go to make up the very ground that we normally walk on, but variety is the spice of life, and a close look at Earth's riches does not disappoint.

In order to read the rock recipes that Earth has to offer, you will need to get to grips with some of the key processes and rock types that are found, as well as ways to identify the differences between rocks, based on the minerals that they are composed of, and the types of crystal forms and structures they display.

In the Beagle-Pod you have access to a vast array of reference material as well as microscopes, hand lenses, and hammers, so you will be able to turn your hand to being the "geologist chef" and mastering the Earth's cookbook.

As you go along you can build a collection of your favourite rocks and minerals as a memento of your travels and your sampling of the local variations that make the Earth special.

The Rock Cycle

Since the formation of the Earth 4.6 billion years ago, it has been made of rocks. These rocks can be quite different to one another, and through millions of years they can be transformed again and again. This is known as the rock cycle: processes such as weathering, melting, and mountain building can all help change rock from one form to another.

The first rocks to form would have been igneous rocks that crystallized from the molten proto-Earth. These can then be weathered and broken down by wind, rain, and chemical reactions to form the particles that collectively make up sedimentary rocks. As they are deposited by rivers and oceans these build up the layers of sedimentary rocks.

The moving plates can then collide to form mountains which bury and change the sediments, through pressure and temperature, to form metamorphic rocks. Here the rock crystals change and grow to metamorphose into different rock types.

As they get buried further they can be heated up to the point where they melt and the rock cycle is complete, with new magma able to rise to the surface and make brand new igneous rocks at volcanoes, ready to be weathered and eroded again and so on. There are clearly many complications to this, but the basic three rock types of igneous, sedimentary, and metamorphic will all be available for you to explore on your journey.

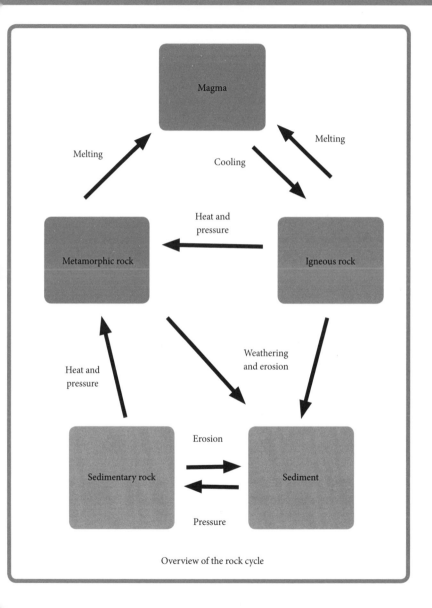

Overview of the rock cycle

Grown from Molten Magma

Igneous rocks, which form from hot molten magma, can be found at depth as well as at the Earth's surface. Magma contains many elements including gases, and can create interlocking crystals as it cools, and even explode if the gas pressure gets too high as it rises to the Earth's surface.

Within the Earth, the igneous rocks form from batches of magma that stall and crystallize within the crust, or that melt where they are. These can represent fossil magma chambers beneath volcanoes, or the plumbing routes of the magma through the crust to the surface, through a network of sheets known as sills and dykes.

They can also be found at the roots of mountain systems, where melting of the rocks creates large granite constructions called batholiths, and make up the base of the oceanic crust.

When magma reaches the surface it can erupt as lava or it can explode and form pyroclastic rocks from hot fragments of rock and ash. These make up the construction of volcanoes and feed the conveyor belt of the ocean surface, where the lava itself forms pillow shapes underwater. Rocks like obsidian, pumice, and basalt form the volcanic landscape. You may well travel through this if you take a volcanic route down into Earth, but either way igneous rocks will form a large part of your surroundings as you go deeper into the crust and beyond.

View into the crater of Marum Crater in Vanuatu

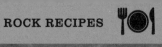

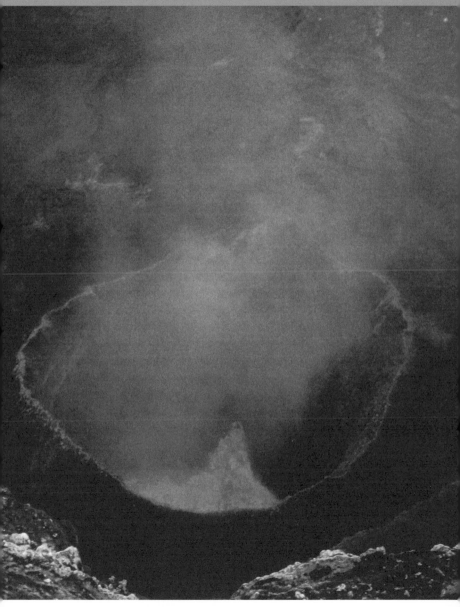

A Sedimentary Layer Cake

Many of the wonderful layers that you will see as you descend into the planet are made from sedimentary rocks. These varied rocks, which are often home to fossils, are made up of broken and weathered pieces of other rocks and shell fragments, and sometimes also contain mineral growths.

Igneous and metamorphic rocks that are exposed in mountains are eroded by the rain, wind, and ice and broken up into their constituent minerals, such as quartz and clays. These are washed along in rivers and deposited on flood plains or into the sea. In these places they go on to form layers of sediments.

In coral beds, the layers of rock are formed by successive generations of coral growths and the precipitation of limestones. These can be formed in tropical waters on the edge of continents or around volcanic islands (atolls).

In extreme examples where complete seas get cut off by the movement of plates, you can have massive evaporation of the water, creating vast layers of rock salt and gypsum.

In the sedimentary layer cake, use your hand lens to explore what grains the rocks are made up of, look for evidence of prehistoric life in fossils, and see how the layers change from those deposited on ancient continents (often with a red hue) to those deposited in the Earth's oldest oceans.

Sedimentary rock layers at Horseshoe Bend on the Colorado River

Bent and Twisted Rocks

With heat and pressure comes change. The immense forces within the planet can change rocks from sediments or igneous form into warped vestiges of their former selves.

These are the metamorphic rocks which are bent and twisted under the pressures of mountain building or subduction, and heated by radioactive decay. They come in a variety of forms and can be close to their previous rock types in appearance or completely altered by the process that forms them. Low to intermediate metamorphic rocks are those which have experienced moderate temperatures and pressures; whereas high grade metamorphic rocks are those which have seen extreme temperatures and pressures.

In the most extreme instances the rocks themselves can start to melt. There are many colorful metamorphic minerals to look for and you can also find folds and twisted rock formations which bear the scars of their metamorphic history.

An example of rocks
that have become
distorted over time

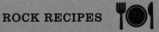

Rocks with names like schist and gneiss, and minerals such as garnet are part of the metamorphic family. You will find such rocks in the roots of mountain chains and along the subduction zones, where, if you are lucky, you might find some blue or green rocks called blue-schists and eclogites. As you travel deeper you can take note of how the rocks are changing and how much more deformed and recrystallized they look as you approach the highest grades where the rocks almost behave like toothpaste.

The Seven Crystal Forms

The wonderful natural shapes of crystals, from the snowflake to the largest crystals on Earth, amaze us even when we are small children. Some of us will have had a crystal garden, where we grew our own crystal shapes in a tank by mixing chemicals and then letting them evaporate.

You may notice that the many crystal forms that you come across as you go down through the rocks of our planet have some

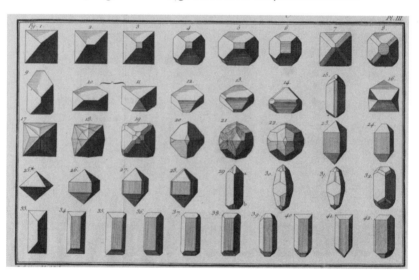

Classic sketches of some common crystal forms by
Jean-Baptiste Louis Romé de l'Isle, 1783

characteristics in common. That is because all minerals belong to one of only seven types, known as crystal systems. Their different types dictate the forms that the minerals take and this in turn leads to their various shapes.

The most simple of the forms is the cubic system (which can also be called isotropic) where many equal-sided shapes can be found. Some classic minerals belong to this group including pyrite, fluorite, rock salt, and diamond. Even gold can occasionally be found in cubic crystal forms. Next comes tetragonal with two sides that are equal and one that is different like a packet of cough sweets. Examples of this form include rutile and zircon. Then there is orthorhombic (for instance topaz, barite, and aragonite), where all three sides are of different lengths like a matchbox.

Then things start to get quite complicated. Monoclinic (gypsum, azurite, and muscovite) has one of its crystal axes at an angle to the others. In the triclinic form all the crystal axes are different lengths and at different angles (not 90 degrees). Examples include albite and kyanite. Trigonal (also known as rhombohedral) forms have three equal axes separated by equal angles that are not right angles and a separate fourth axis (examples include calcite, quartz/amethyst, and corundum – rubies and sapphires). The hexagonal system (such as beryls including emerald and aquamarine) has sixfold symmetry down one of its crystal axes which can make hexagonal shapes.

Don't worry if this all sounds rather complicated: you will have a box of these simple crystal shapes in your Beagle-Pod which you can use to compare the crystals that you find. See if you can see similar shapes and even identify some of the minerals.

How Hard is Hard?

All rocks are made of minerals and all minerals are different from one other. This can be due to their composition and the type of crystal system they belong to, and these in turn help to define the key properties of the minerals. One of these properties is their hardness. There is a scale from soft to hard which defines how strong the minerals are. This is basically measured by testing which minerals are able to scratch ones softer than themselves. The Mohs scale of mineral hardness runs from 1 (softest) to 10 (hardest).

All minerals fall somewhere along this scale and there are some key ones which are used to help define it. You will be familiar with some of these, but some are less well known. Coming in at number 1 is talc, then numbers 2 to 5 are gypsum, calcite, fluorite, and apatite. (It's a kind of reverse scale so that the hardest is actually not 1 but 10). At 6 we have feldspar, then 7 is one of the most common minerals, quartz. Finally the top three are topaz (8), corundum (9), and the hardest of them all, diamond at number 10.

One of the fun things about the scale is that you can use various things to test the hardness of any mineral that you come across; your finger nail (2.5), a copper coin (3.5), a steel nail (5.5), glass plate (6), and a mineral streak plate (close to a hardness of 7). You will find these along with the models of crystal shapes in your Beagle-Pod science kit.

Tales from the
Time Machine

*Life can only be understood backwards; but it
must be lived forwards.*

Søren Kierkegaard

The Layers of Time

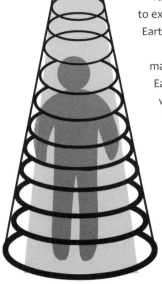

You would think that as you travel down within the Earth the rocks would get older, and for the most part that is correct: it's certainly the logical assumption to make. If new rocks are formed at the surface, then as they get buried by newer rocks they become older and so on. However, due to the way the Earth's plates move slowly through time, they can crash together and cause older rocks to be moved above younger rocks, so in some cases the tables can be turned.

Tales from the Time Machine is your chance to explore the layers and structures within the Earth that record its geological history.

It's like using your Beagle-Pod as a time machine to go back and see what was on Earth at key moments in its history, and with over 4.6 billion years at your fingertips there is a lot out there.

Many travellers have described this part of the journey as the most rewarding, as the sheer variety of rocks, fossils, and spectacles draw people back to their childhood days or to the first time they became fascinated by the natural world around them. In this section you will get a flavour of some of the things the time machine

The layers of rock are the equivalent of layers of time in the planet's history

can offer. So sit back, tune in the dial on your Beagle-Pod to a time in Earth's past, and you will see a 3D map of places you can explore within the planet and where rocks of around that age exist.

What are you waiting for, go take a look around!

The Geological Record

The 4.6 billion years to the present day, have seen a lot of change within the Earth. This vast amount of time is captured and split up into a chart known as the geological record.

It has generally evolved to accommodate key changes that scientists have observed over the years that in turn relate to major changes within the Earth. Terms such as Pre-Cambrian, Palaeozoic, Mesozoic, and Cenozoic should start to become familiar to you during your travels, as well as more specific time periods such as the Jurassic or Cretaceous.

The latter part of the chart is designed around the evolution of life from the great Cambrian Explosion (or Cambrian Radiation) to some of the key events such as the End-Permian extinction.

Your Beagle-Pod has the geological record as a time chart which is constantly updated as you travel around and within the Earth's crust, mantle, and core.

You can see exactly which time period the rocks you are travelling through are from. This will help with your exploration of different geological time zones, and will also enable you to find key rock types or to look for various types of fossil treasure.

An overview of the geological record as we travel back through time

Cenozoic Era

Quaternary Pleistocene/ Holocene 2.5 million years ago	Neogene (Miocene/Pliocene) 20 million years ago	Paleogene (Paleocene/Eocene/ Oligocene) 66 million years ago

Mesozoic Era

Cretaceous 145 million years ago	Jurassic 201 million years ago	Triassic 252 million years ago

Paleozoic Era

Permian 289 mya	Carboniferous 358 mya	Devonian 419 mya
Silurian 443 mya	Ordovician 485 mya	Cambrian 541 mya

Proterozoic Eon

Neoproterozoic Era 1200 mya	Mesoproterozoic 1600 mya	Paleoproterozoic 250 mya

The Oldest Rocks

To find the oldest rocks on Earth, you need to go to the parts of the crust that have survived the constant movement, crashing, and destruction that the plate-tectonic engine has unleashed on Earth over millions and millions of years.

These are parts of the Earth's continental crust known as the "cratons." They have been more or less stable since Pre-Cambrian times, and are found within the major continental plates: Africa, South America, North America, Eurasia, and Australia have notable examples. The oldest rocks on Earth are found in the cratons and are usually dated using uranium/lead decay in minerals like zircons, which are extremely robust and reliable recorders of geological age.

The oldest rocks at around 4.031 billion years old are found in the Slave craton in northwestern Canada. They are examples of a type of metamorphic rock known as gneiss. The oldest fragment of terrestrial Earth has been dated as a zircon grain from ancient sediments that were deposited around 3 billion years ago. The grain itself was eroded from a rock that gives an age of 4.404 billion years.

You will probably be heading to some of the cratons in search of diamonds anyway, but keep an eye on your geological-time dial, as you never know just how far back in time your Beagle-Pod will take you, and you just might find an even older piece of the Earth.

An exposed section of craton in Michigan, showing ancient glacially-smoothed greenstone (metamorphosed basaltic pillow lava)

Trail of the Ice Ages

To trace the past it is fascinating to drill down through the massive ice sheets of Greenland or Antarctica: the latter could be done as part of a visit to Lake Vostok.

As you go down through the layers you effectively go back in time so you can find information about the Earth's climate changes over the years. One of the longest ice cores in Antarctica has revealed a climate record going back some 800,000 years, and it is thought that you can even get ice as old as 1.5 million years way down in the deepest ice of Antarctica.

This reveals evidence that the climate of the Earth has been through warmer and cooler periods. The colder ones are called glacial periods, the warmer ones interglacials. But what about the

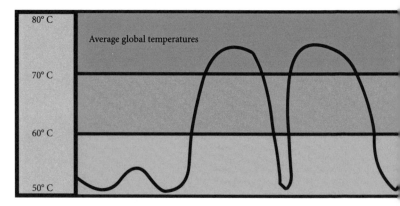

2.4 billion years ago

Ice Ages? Well, we are currently in one and there have been at least five major Ice Ages in the Earth's past.

For more insight into the icy past of Earth beyond the oldest periods you will find preserved in the ice, you will need to look for special clues within the rock record.

It's not just the U-shaped valleys and drumlin landscapes on the surface that provide evidence of cooler times on the planet. Some types of sediments can reveal Ice Ages going way back in the rock record. Here you need to look for specific clues such as random "drop stones" (large boulders lying in deep water sediments that have dropped from rafted ice in the ancient seas).

Alternatively you might spot vast glacial sediments occurring in rocks that have formed closer to the equator than you would normally get ice. All of these are clues that you can look for as you go through the geological past beyond the ice sheets themselves. And then beyond them all, there is the biggest Ice Age of them all, when the entire Earth was a snowball . . .

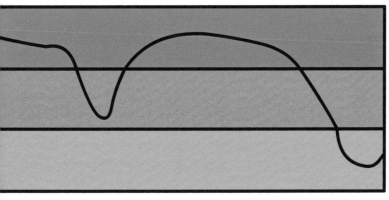

Today

When the Earth was a Snowball

If you think it's cold in the winter, you can always nip off to hotter climes and get some valuable rays of sunshine. However, there have been times in Earth's dim and distant past when it was winter everywhere. Imagine icy wastes as far as the eye can see, even down to the equator, in a time known as "Snowball Earth."

The Earth has been significantly colder than it is now for large periods of time, known as Ice Ages (not to be confused with glacial periods). As we have seen, there have been five major Ice Ages on Earth. These are: the Huronian (2.4-2.1 billion years ago); the Cryogenian (approximately 850 to 630 million years ago); the Andean-Saharan (approximately 460 to 420 million years ago); the Karoo (approximately 360 to 260 million years ago); and the

current Quaternary glaciation (that started around 2.6 million years ago and is continuing today). It's the Cryogenian Ice Age that is thought to have been one of the biggest and that has coined the term "Snowball Earth." This is because it is believed that the vast ice sheets of the Cryogenian actually reached the equator, turning the Earth into one giant snowball.

Although you can't see the ice today, you can drill through the rocks of this age in the Beagle-Pod and look for the clues. Ancient glacial sediments are called "diamictites" and are very poorly sorted continental sediments (called glacial till in the present-day examples), and many are found in the geological record of the Cryogenian. There might be rocks that show "striations," scrapes and scars left by the moving ice, and you may even be lucky enough to find a drop stone in deep water sediments.

If you find some of these clues and the pod says you are in the right time period (where there are rocks from 850 to 630 million old), you may well be witness to the coldest period ever witnessed on the planet, Snowball Earth.

The Earth's LIPs

Periodically through history, the Earth has experienced huge volcanic events, the like of which we haven't witnessed in our time. The mixture of a convecting planet and the rifting and moving continents means that every now and then a burst of hot material rises from deep within the Earth and strikes its surface creating what are known as "large igneous provinces" or LIPs.

Gigantic outpourings of lava and gases are associated with the LIPs, which can strike in a relatively short period of geological time (a million years or less) causing stresses to our planet. LIP events have been associated with some of the major Earth crises, and were responsible for shaping the course of the planet's evolution for the last few hundred million years or so.

They are thought to be mainly caused by what are known as "mantle plumes". These uprising mantle plumes bring hot and fertile mantle material that is primed for melting up towards the Earth's surface. The eventual result is the rapid melting and volcanic eruptions that form a LIP.

These gigantic Earth events have left big scars on the planet, where vast volcanic rock units can be found and explored. A map of the Earth's LIPs overlaps frequently with volcanic areas. Some key LIPs you can explore with the Beagle-Pod include the Deccan Traps which formed 66 million years ago, and the Siberian Traps which erupted some 252 million years back.

Earth's Lifeline

Heaven is under our feet as well as over our heads.

Henry David Thoreau

Earth as a 24-hour Clock

Set our stopwatch, we are on a journey through Earth's history as though it were a 24-hour clock. From 4.6 billion years to the present day, things are going to move fast . . . or are they?

Ready, set, go: . . . so, 00:10 and we see the formation of the Moon, a pretty big event. And we see the origin of life and the oldest fossils between 04:00 and 05:30.

Mmmmmm . . . not much is going on at this point, I think it's time for bed. Let's sleep through the atmosphere becoming oxygenated at about 11:00 and put the clock on snooze to wake

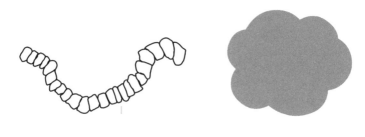

First plant fossils 04.30

Atmosphere becomes oxygenated

Some highlights in Earth's timeline

us up when life really gets going in the Cambrian Explosion.

Well, that was a long snooze, as it is now around 21:00 and the alarm finally goes, with the Trilobite Derby (see next page) due to start at around 21:10, followed by the first insects.

Dinosaurs finally join the party at around 22:56, but they don't stay for long: they disappear around 23:40, hardly a big showing. And it's just before the midnight hour (approximately 23:59) when the first human-like animals (which are collectively known as hominids) appear.

Whoooosh and that's it! We really have not been here for a long time when you think about it this way, but it's all the time that is locked up in the world that you can see on your journey to the centre of the Earth, and don't worry, that's definitely going to take a lot longer than 24 hours.

24.00

Trilobite derby 21.10

Dinosaurs 22.56

Hominids 23.59

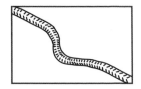

The Trilobite Derby

Some of the most diverse and iconic critters that ever crawled the Earth, the trilobites were a group of animals that blossomed near the start of the Cambrian (at around 521 million years ago) at a time known as the Cambrian Explosion (or Cambrian Radiation). This was a special time in evolution when most major animal phyla appeared. These segmented beasts, which look like a cross between a woodlouse and a king crab, made the ocean floor their home. They were robust creatures, as they lasted, in one shape or form, for nearly 300 million years. They were finally wiped out at the end of the Permian (which was approximately 252 million years ago).

They were at their most widespread during the Cambrian and Ordovician times, during which the ocean floors were teeming with trilobites cruising around like some kind of race or "Trilobite Derby." Their fossil bodies are the exact picture of a creepy-crawly, with a head (cephalon), a body (thorax), and a tail (pygidium). Some had spikes and some had great big compound eyes with many small lenses. Others had weird spines and antennae, and their bodies sometimes hid some wiggly legs underneath.

As you explore the fossil beds of the Cambrian and Ordovician layers, you will sometimes find a strange trace fossil, a bit like dinosaur footprints but much smaller. That is because all the evidence that is sometimes left of a trilobite is a fossil track left by its bustling feet, known affectionately as Cruziana (pictured above).

Paraceraurus trilobite found at the Volchow River in Russia

The Coral Chasms

It's not always the big bugs and animals that take the limelight, it can sometimes be the little ones that make a big splash. An example of small-makes-big includes the formation of chalk, of the sort you see at the white cliffs of Dover and which you can often cut through in rocks which date to the Cretaceous period.

The chalk is made up of tiny fossil bodies from marine algae, called coccoliths. Another great example is that of the corals. Anyone who has dived the Great Barrier Reef will understand that over time these small coral creatures can construct vast city-like structures of carbonate reefs.

In the ancient rock record, corals have also left their mark, building great fossil reefs that can be discovered in limestone rocks, particularly in the Carboniferous times (between 363 and 325 million years ago), as well as good examples in the Jurassic and Cretaceous. The soft carbonate limestone rocks are easily eroded and dissolved in ground waters and can form vast cave networks as a result, so you will sometimes find great coral chasms, rich in these fossil reefs. Many other life forms are found around reefs, and this was also the case in the past, which means that many a fossil can be found here, so keep your identification book and hand lens at the ready.

Coral on the Great Barrier Reef

Day of the Dinosaurs

The child in all of us is fascinated by dinosaurs, and the thought of roaming around at a time when these giants of Earth were alive has been the inspiration for many a film and book. Although we cannot get to see these scaly wonders in life, we can go through the

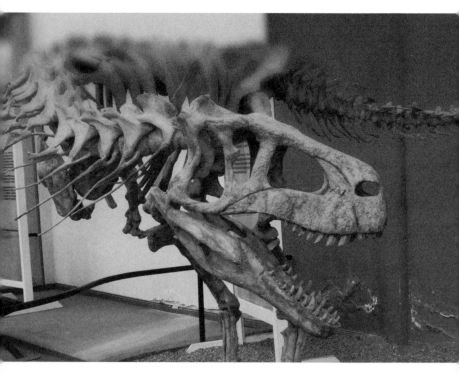

rock units that formed when they did roam the Earth and look for evidence of them. This can be found in the fossil remnants of a past world that mark the day of the dinosaurs.

For 170 odd million years the dinosaurs roamed the surface and swam in the oceans, and some even flew in the skies. They ruled our planet as some of the largest and also smallest birdlike reptilian hybrids until their final demise about 66 million years ago.

To see the remains of these wonderful beasts you will need to program the Beagle-Pod to aim for rocks from the Triassic, Jurassic or Cretaceous. Popular dinosaurs include Diplodocus (age ~152 mya),

Triceratops (age ~68 mya) and of course the big daddy himself, Tyrannosaurus Rex (age ~68 mya). Since the *Jurassic Park* series, some of the smaller ones such as Velociraptor (age ~75 mya) have also become more widely known.

If you are lucky you will find bones, but also look out for trace fossils such as dinosaur footprints in ancient sandstones. If you are really lucky you may come across a dinosaur nest with the eggs still present, and for the more squeamish you could even find a fossil known as a coprolite . . . fossilized poo!

Face to face with a dinosaur

Earth Crisis

The Earth has not always been a good place to live. At some stages of its long history there have been major events that have caused catastrophic changes in the climate and resulted in the deaths of groups of species at the same geological time.

These "Earth crises" result in what is known as a mass extinction event. Several of these have occurred on Earth since life species have been recognized in the rock record. But what caused them?

Ordovician–Silurian extinction
Around 439 million years ago, 86% of life on Earth was wiped out due to glaciation and falling sea levels, possibly because of too many plants producing carbon dioxide.

Late Devonian extinction
Around 75% of species were lost around 364 million years ago either in one single major event or over a long period. Trilobites were almost eliminated along with many others.

Permian–Triassic extinction
Also known as the End-Permian extinction, this happened 252 million years ago, possibly due to volcanic activity. It is seen as the worst in history: only 10% or less of species survived.

One of the main groups of protagonists in this climate cocktail, is our volcanic friends, or in this case enemies. Every now and then the interior of the Earth unleashes a hot spot or plume that can heat the upper parts of the crust and mantle like a blow torch. This excess heat can result in huge volumes of magma being generated and erupting in a geologically short period, called large igneous provinces (LIPs). The result is a wave of gases that add to and change the atmosphere and cause an Earth crisis. One of the most widely recognized of these is the End-Permian extinction event. You can aim to visit the rocks that span this and other events as you journey through the planet. Sometimes the picture is even more complicated, for instance when both volcanoes and other factors, like a comet's impact, all go into the Earth crisis mix.

Triassic–Jurassic extinction

This occurred between 199 million and 214 million years ago. Many species died out due to volcanic activity, climate change and also possibly an asteroid impact.

Cretaceous–Paleogene extinction

The best known of the Big Five, this led to the extinction of the dinosaurs. Volcanic activity, asteroid impact, and climate change ended 76% of species, 66 million years ago.

Are We Next?

The big question is whether or not the current Holocene era is heading into a sixth mass extinction. Human activity and climate change have already led to many species being lost.

A Siberian Tale of Disaster

Never before during the times when significant life was roaming the Earth have we seen a disaster like that experienced by the Earth at the end of the Permian. A catalogue of events conspired to trigger the most deadly mass extinction on the planet. You can explore the rocks that helped make this happen as well as the death zone in the sediments which record this time period. Tune your Beagle-Pod to 252 million years ago and add the keyword Siberia, and you will be taken off to the basins around the Siberian Plateau which hold the evidence for this catastrophic event, which is known as the End-Permian extinction.

First, you can drill through massive lava flows and underground igneous intrusions within the sediments that mark a major volcanic event that is known as the Siberian Traps: this was one of Earth's largest ever igneous provinces.

You may also spot strange pipe structures at the tips of these intrusions which pierce up to the surface. These mark the sites where many tons of toxic gases, generated when the intrusions boiled, combined with the gases from the volcanic eruptions to cause a massive climate crisis.

When it was finished, over 90% of marine species and around 70% of land dwellers had been killed off. This is the most striking example of a mass extinction that was probably caused by a major volcanic episode. Take a further eerie tour of the sediment sequences that host this End-Permian layer and, as you traverse the boundary, you will see how the life forms disappear.

Giant basalt eruptions like this one in Iceland would have fed
the many lava flows of the Siberian Traps

Where Have All The Dinosaurs Gone?

A famous extinction event occurred on the planet around 66 million years ago which saw the death of the dinosaurs, ammonites, and a number of other species. Vast outpourings of lava consumed the landscape in India around this time resulting in the thick and extensive set of rocks called the Deccan Traps.

As with other large igneous provinces, the Deccan has been implicated in this Earth crisis. However, this is a crisis with a twist, because there was also a comet impact which has been linked to the same mass extinction event.

The jury is still out as to whether the volcanoes or the comet led to the demise of the dinosaurs and other lifeforms at the End-Cretaceous extinction. It is most likely to have been the coincidence of both that placed too much stress on the Earth's climate.

The thick layers of volcanic rocks that make up the Deccan make for a great excursion, as you can travel through the smoking gun which marks the time in Earth-space when this extinction event actually happened. Maybe you can even find some of the last dinosaur eggs lost within the earliest eruptions of the Deccan.

You can also hunt for the famous iridium spike that marks the moment when the comet hit, as this layer can also be found within the lava flows, suggesting that both of these catastrophic events were very closely spaced in Earth's history. When you come out from the top of the last flow you may well be left pondering "where have all the dinosaurs gone?"

The Western Ghats mountain range, as seen from Mahabaleshwar,
India, show the characteristics "step" shape of traps

Spotting Earth's Survivors

There are a very small number of extraordinary beasts on the planet today that have been around for a long, long time. Some of them have been around for over a million yearsat least, their ancestors have – without much evolutionary change. This band of survivors are so perfectly adapted that they have survived catastrophes where others have fallen.

Four examples you can follow back through time are the sea shell nautilus, the dragonfly, stromatolites, and Jaws himself (the shark, pictured right). The youngest of the four, the dragonfly, dates back to around 315 million years ago, with dragonfly fossils having been found in the ancient swamp rocks of the Carboniferous period.

The shark goes back some 405 million years (which would make for a pretty long movie). Nautiloids go back even further still to the late Cambrian time period, just over 500 million years ago.

The oldest of the lot are the stromatolites, some 3.7 billion years old, as shallow water microbial mats were some of the earliest lifeforms on the planet. Tune in the Beagle-Pod to the oldest time for each species and see if you can find some of the earliest examples of these resilient survivors.

In The Deep

Wherever he saw a hole he always wanted to know the depth of it. To him this was important.

Jules Verne

The Low Velocity Zone

The low velocity zone is found close to the boundary between the parts of the upper mantle that make up the lithosphere and the asthenosphere, around 50-65 km (31-40 miles) below the ocean spreading centres and over 120 km (75 miles) beneath the older oceanic crust. Its main characteristic is its unusually low seismic shear wave velocity compared to its surroundings. This means that seismic P waves are somewhat slowed and S waves are slowed as well as being partially absorbed.

Why so slow? Well this zone is somewhat special as it contains very low volumes of melt: small sleeves of magma lie between the crystal grains in this part of the mantle, which explains why the seismic waves vary here.

If these melts are tapped into, as they could be at mid-ocean ridges, you can feed the magma up to the Earth's surface. So in this way, as you travel along the low velocity zones you are moving through one of the Earth's interior layers where melt is generated and, possibly, where some volcanoes are made.

In truth, the circumstances in which volcanoes are born require special conditions. However, what is clear is that these conditions are likely to be found in the low velocity zone.

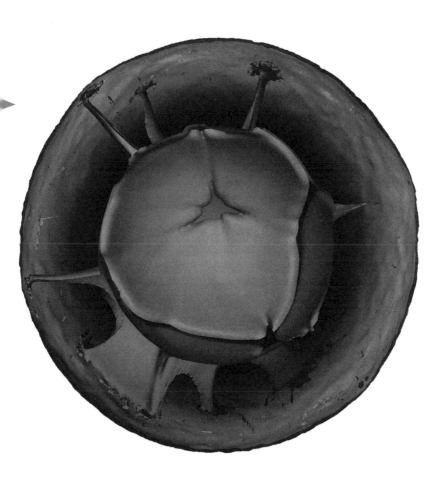

A digital representation of the Earth's core with plumes reaching up
through the mantle to the surface (see next page)

Hot-Spotting Your Plumes

As you drive the Beagle-Pod through the Earth's mantle, you will have the opportunity to see some of the deepest expressions of the exchange of material within the planet. Structures known as plumes mark the slow rise of anomalously hot, and sometimes compositionally different, parts of the mantle.

The surface expression of these features is often called a "hot spot." They can be very obvious when found on the ocean crust.

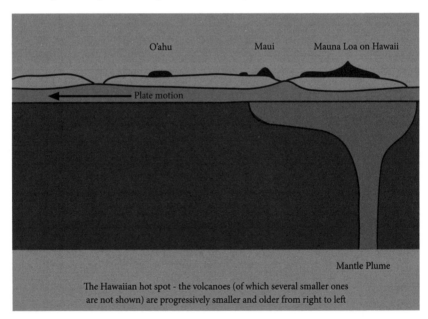

The Hawaiian hot spot - the volcanoes (of which several smaller ones are not shown) are progressively smaller and older from right to left

They represent spots on the planet's surface where the upwellings have caused lots of melt, and extra volcanic activity. The structures tend to be roughly cylindrical and can rise from extremely deep in the mantle, even as deep as the core-mantle boundary.

The classic trail of a plume impact on the ocean crust is that of the Hawaiian hot spot. As the Pacific Plate moves across this plume structure, the hot plume (which is a bit like a blow torch in the mantle) under Hawaii causes excess melting and the rise of one of the largest volcanoes on the planet (Mauna Loa), making up the Big Island. The movement of the plate has, however, resulted in what is known as a hot spot trail, as the successive volcanic centres have been moved away from the plume with the motion of the plate. A view of the islands and the sea floor around Hawaii reveals this chain, not only as the islands but as submerged seamounts that trail away from the currently active hot spot at the plume's centre. It is great to explore these swirling structures within the mantle: they also make for popular return routes to the surface.

Surfing the Core-Mantle Boundary

One of the most important layers within the depths of the Earth is the core-mantle boundary (CMB). This marks the zone where we go from solid silicate mantle into a liquid material that is rich in iron and nickel.

The boundary itself is named the "Gutenberg discontinuity" after the German scientist Beno Gutenberg (1889-1960) who spent a great deal of time trying to determine its location.

As you approach this boundary you will notice that things start to change, and that the mantle becomes quite different in behaviour. This is the D" layer (see p.30) in the lowermost part of the mantle, and you should expect to start to experience the difference within about 200 km (125 miles) of reaching the core-mantle boundary, unless you are near the large anomalies known as superswells that are found beneath the Pacific and Africa (see the following two pages).

Complex mineralogical changes occur just above and around the CMB, and it's not entirely clear what the true nature of the contact between the solid and liquid areas is, with small scale structures that occur along the boundary being known as ultra-low velocity zones.

The actual surface of the layer is not a simple sphere because of these complications. Indeed, the very nature of these differences may help contribute to and maintain our magnetic field. You can spend some time surfing along the liquid top of the outer core and explore this fascinating zone of the deep Earth.

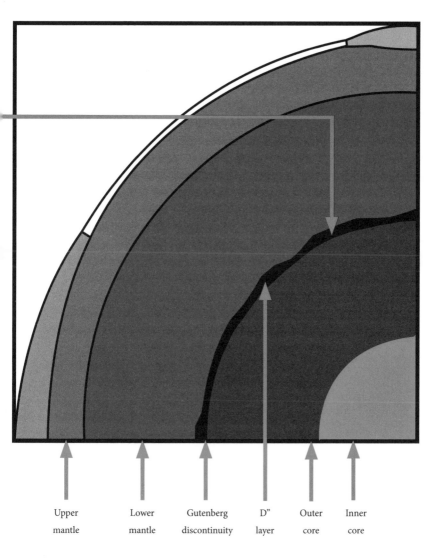

| Upper mantle | Lower mantle | Gutenberg discontinuity | D" layer | Outer core | Inner core |

An overview of the core-mantle boundary

The African & South Pacific Superswells

Within the deepest layers of the mantle there are some gigantic anomalies that go almost unnoticed, but which the wine growers of Southern Africa are quite happy about. This is because two particular vast regions, one beneath the Pacific and one beneath Africa, have a direct effect on our planet's surface.

These zones are known as large-low-shear-velocity provinces (LLSVPs), They equate to anomalously low velocity parts of the base of the mantle that rise up significantly from the core-mantle boundary. You can feel this change as you drive your pod through into these areas (a sort of extension of the D" layer) as the

Cartoons of the locations of the African and South Pacific superswells, known respectively as "TUZO" and "JASON"

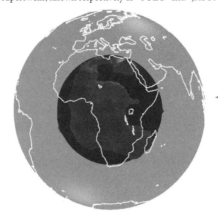

Beagle-Pod reacts to its surroundings.

We like to put things into boxes, as we do when we categorize the Earth into its key layers. However, the Earth system communicates in a more complex way across the layers, from core to surface and back again. A simplified map of the Earth's interior shows the LLSVPs marking the main zones of upwelling in between areas where we get mainly subduction. As a result, the planet has been described as a two degree Earth (a pattern of two downwellings which have high velocity and two upwellings each of which have low-velocity). These vast LLSVPs lead to elevated regions on the Earth's surface above, which are called the African superswell and the South Pacific superswell. Africa has a peculiarly high elevation which can be tracked around it, particularly in the southern and eastern plateaux of the continent. LLSVPs have also been implicated in the birth of hot mantle plumes. However, it's on the modern-day elevated slopes of the southern Cape, where we benefit directly from them. So as you sip your South African wine, don't forget to raise a glass to the African Superswell.

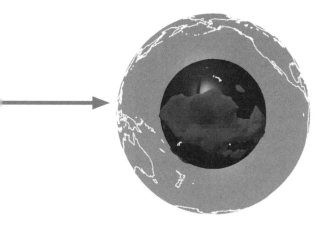

The World On Your Shoulders

The Beagle-Pod will shudder to a halt, buzz like a vibrating phone and then there will be silence. It is programmed to do this. Your dials will all light up, and a big red beacon will blink on and off above your cockpit. Then, after a few moments, your chosen voice will state: "Welcome to the Centre of the Earth, we hope you have enjoyed your journey." (Remember you can choose from a variety of voices including famous people, just like the GPS in your car.)

You can read all the information about your route to the centre as the Beagle-Pod prints off your Centre Earth Travelogue summary. This is the first half of your full travelogue (which includes your journey back). Then eerie silence as you gaze at your sphere compass, swirling around in abject confusion.

Like Atlas himself, everything above your head is all that makes up the Earth. You have finally done it, you are at the CENTRE OF THE EARTH.

Take a few moments to reflect on your achievement. Everything that you have ever known is above you, and from this point you can head up to anywhere on the planet.

You should already have chosen your route back by this stage, so you will need to program this into the Beagle-Pod if you have not already done so. Remember that if you have missed anything on the way down you have one more chance to see it on the way back before you pop out at your chosen return destination.

Well done and welcome to the world of the Earth Wormers!

The Return Journey

There is nothing like returning to a place that remains unchanged to find the ways in which you yourself have altered.

Nelson Mandela

Visualization of upwelling mantle structures from the core-mantle boundary. The Iceland plume (the dotted vertical line) can be seen currently centred near the Öræfajökull volcano.

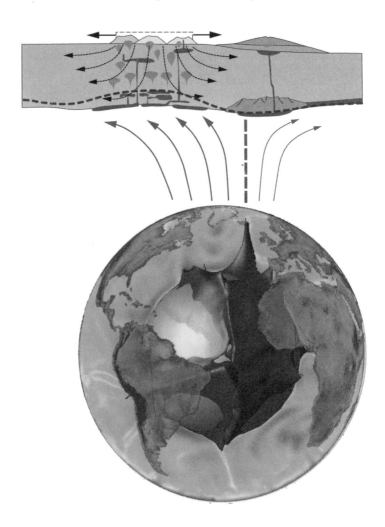

Riding The Mantle Plume

To get back home from the centre of the Earth you can aim for pretty much anywhere on the surface of the globe. You have a 360 degree choice of where you want to pop up, which may depend on where you live or whether you want to take some extra vacation time before you finish your trip. In order to facilitate your journey to the surface you can take advantage of some of the Earth's internal structures and movements to get you home.

Mantle plumes are hot upwellings of mantle material that rise up from deep within the Earth. Some famously volatile places such as Iceland and Hawaii sit on top of such structures, which helps to explain their volcanic origins. These upwellings have been imaged down to the core-mantle boundary in places, but in reality they move very slowly as the Earth's mantle convects and cools, forming fantastic 3D blobs, tubes, and shapes within the mantle. Coming up a mantle plume has the added advantage of allowing you to travel through the inside of a volcano.

You can ride the mantle plume wave up to the surface and pop out at Hawaii. Or why not try what we call the "Reverse Jules Verne," and take the Iceland plume to the surface and have a day or two exploring the volcanoes and icecaps in Iceland?

This is the site that Verne had as the entry point to his mystical novel *Journey to the Centre of the Earth*, so what more fitting a place could there be to end your own adventure?

Opposites Attract: the 180° Way

A popular way to exit from the centre of the Earth is to pop up on the opposite side of the Earth from where you live. People do this to see what is on the other side and to have a sort of post-expedition wind-down. You need to be somewhat careful as, depending on where you live, the exact opposite part of the Earth may be underwater in the middle of an ocean.

When this is the case most people chose to pick the closest land mass to the point exactly 180° from home. Others have a yacht waiting with a gin & tonic for their arrival from deep within the Earth. You can enjoy a few days holiday on the waves, before taking your final flight back home.

The location that is directly on the opposite side of the Earth to a paritcular location is known as its "antipode." If you wanted to take this approach and you started from Southern America, you would pop out around China, the Philippines, or Indonesia. Many parts of Africa are antipodal to islands in the Pacific. And from Spain or Portugal you would end up in New Zealand.

For many other starting locations, you will need to research small islands nearby or have a boat waiting for you at your antipodal arrival point. This is because only approximately 15% of Earth's land is antipodal to other land.

A visualization of the areas where antipodal points are
both on the land (darker grey) rather than in the ocean.

Pick your Favourite Landmark

For many, the journey back up to the surface provides the ideal opportunity to emerge at a landmark that they especially revere, whether it be an iconic building, a wondrous landscape or a specific location that is special to them. Each person has their own favourite, and the magic of the pod means that you can pick just about anywhere to pop out.

Your location may be spiritual, like the Taj Mahal or Mecca, it may follow some engineering feat such as the Eiffel Tower or the Empire State Building, or any landmark you would love to visit.

Popular geographical wonders include Uluru (formerly Ayers

Uluru (formerly known as Ayers Rock)

Rock), giant waterfalls (e.g. Victoria Falls in Africa; Angel Falls in South America; Niagara Falls in North America), great mountains (Mount Everest, Kilimanjaro, The Matterhorn), or just a wonderful landscape to end up in like the Namib Desert, the Painted Desert, the islands of Japan, or the fjords of Norway.

Each to their own, but just remember some of the more popular sites may need to be booked and planned in advance, and don't forget to think about how you are going to get home once you pop out in your favourite landmark. For remote locations such as Uluru, it is especially important to plan in advance as you don't want to add a wilderness trek to your holiday.

Up and Then Back in Time Again

If you are a true adrenalin seeker then this popular ending is one that will have you gripping the seats of the Beagle-Pod. Having travelled through the layers of time within the Earth, you pop up at the one place where you can travel through almost half the Earth's history in one wild river ride.

We are talking about the Grand Canyon, and the final journey of you and your pod through the 430 odd kilometers (270 miles) of foaming white water that makes up the most spectacular natural gorge in the world. You will see rocks from close to the present day and back through to the Pre-Cambrian on this final journey, as you and the pod bob along. There are wonderful sections through the gorge, named by the first expedition down it in 1869, which reflect the rocks and scenery you will be passing.

Enter the Marble Canyon and marvel at the Redwall Cavern. Dare to pass the gateway to the Granite Gorge and be tied up in the Granite Prison, and finally be bold enough to ride the "Lava Falls" where a monstrous rapid has been formed where a lava flow dropped into the gorge some 100,000 years ago. In total you will have travelled through 80 big water rapids, and will have travelled back in time nearly 2 billion years, this time on the surface of the Earth.

The Grand Canyon

In the Back Garden

It's a very special journey you are undertaking, and no doubt your friends, family, and local community are all backing you to make a successful expedition down and back up again. Why not arrange to arrive back home in your own garden or village centre?

This is particularly popular with those who have received a

generous amount local sponsorship to set up their journey and those who are raising funds for a special cause. Arriving back in a fanfare of banners and bunting makes for a very rewarding and memorable way to end your trip.

You will need to get the local permissions to pop out of the ground if you are using your village square or a specific local monument, but these can be arranged as part of your arrival party. Also there is an extra clean-up plan than needs to be put in place to fill up the hole you will emerge from. Luckily the Beagle-Pod is designed to leave minimal damage to the surrounding ground, and for many who are arriving in their own property, leaving a small exit crater as a landmark in the garden will provide a great talking point over a barbecue for many years to come.

Your Trophy - The Gertrude Award

A major achievement deserves some kind of certificate, placard or trophy: a way of celebrating the event, and having a lasting aide-de-memoire that you can show off to future generations. Your journey to the centre of the Earth is no exception, and each person that makes this journey, henceforth known affectionately as a "Human Earth Worm," will receive a trophy to commemorate their exceptional feat. You will be awarded this at your completion point, where you arrive back to the Earth's surface. This will need to be noted before you head off so that arrangements can be made for the ceremony.

Although she was not in the original book, Gertrude the Duck

became the star of the show in the 1959 film adaptation of Jules Verne's *Journey to the Centre of the Earth*. She is one of the few animals to have their own entry on the IMDb actors guild, and made a big splash as part of this iconic film. As a fitting tribute to Gertrude and to your fantastic achievement as a Human Earth-Worm, the "Gertrude Award" is the trophy that will adorn your mantelpiece as a testament to your achievement. Not many people have a Gertrude Award as it is only a select few that have been on the incredible journey to the centre of the Earth.

Welcome to the elite club of the Human Earth-Wormers, and remember to share the findings of your travel with others.* Life is about the stories that paint a picture of your world to others, and the best yarns always come from the mouths of those who have experienced things at first hand.

*As part of sharing your experience, your opinion really matters to us. Once you have completed your journey and received the Gertrude Award, you will be given a unique code that links into our web portal so that you can give your feedback on the journey to the centre of the Earth. You could also place reviews on popular travel sites, and hopefully inspire others to take the deep plunge.

Index

Picture credits

All illustrations and images by Diane Law except for the following:
Tunnelling machine (p.6): cooper.ch via Creative Commons Basalts (p.14):
Ashley Dace/CC / Cross-section (p.9): Dreamstime / Earth (p.12, 49):
Dreamstime / 3D Earth models (pp.16, 18, 19, 137): courtesy of Fabio Crameri,
Centre of Earth Evolution and Dynamics (CEED), University of Oslo / Section
of Mantle (page 17): Adapted from USGS / Plate Map (page 24): adapted
from USGS / Oldham and Lehmann (p.29): Public domain / Earth Dynamo
Models (p.33): Dr. Gary A. Glatzmaier - Los Alamos National Laboratory-
UCSC-NASA / Snider-Pellegrini Weneger map (p.35): USGS / Ridges (pp.36-7):
Shutterstock / *Trieste* (p.44): Public domain / Chimborazo (p.47) / Francesco
Ballo / Atmosphere (p.51): NASA / Dallol (p.52): Waltatekie/CC / Continent
Maps (p.54, 55): NASA / Diamonds (p.57): Swamibu/CC / Fossils (p.59):
Richard Wheeler/CC / Silfra (p.60): Diego Delso / Photos (pp.61, 64, 69,
103, 126, 131, 133, 153): Supplied by Dougal Jerram / Rift Map (p.62): NASA
Hercules (p.63): NOAA Picture Library/Public domain / Artistic Impression
of Lake Vostok (p.67): Nicolle Rager-Fuller, NSF / Earth's Magnetic Field
(p.79): adapted from NASA / Lava (p.81): Hawaii Volcano Observatory (DAS)
/ Great Comet (p.83): E.Weiss / Chile (p.86): Claudio Núñez / Magma (p.88):
Shutterstock / Sun (p.91): NASA/SDO / Pebbles (p.95): Sean the Spook/CC
/ Marum (p.99): Geophile71/CC / Horseshoe Bend (p.101): Paul Hermans/
CC / Crystal Models by Jean-Baptiste Louis Romé de l'Isle, 1783 (p.104):
Collection Teylers Museum, Haarlem (the Netherlands) / Mountain (p.109):
Shutterstock / Craton (p.113): James St John/CC / Trilobite (p.123): Vassil/CC
/ Coral (p.125): Toby Hudson/CC / Shark (p.134): Mark Conlin, SWFSC Large
Pelagics Program / Volcano (p.139) G.E. Ulrich, USGS / Superswells (pp.142-3):
Sanne.cottar/CC / Iceland Plume Figure (p.146): courtesy of Trond Torsvik,
CEED, University of Oslo / Map of antipodes of the Earth p.149): in Lambert
Azimuthal Equal-Area projection (by Citynoise, Wikimedia commons) / Uluru
(p.150): Stuart Edwards / Champagne (p.154): Shutterstock